in Bytes we tr@v@l

sponsored by

Sun
microsystems

in Bytes we tr@$_v$@l

Lin Hsin Hsin

World Scientific
Singapore • New Jersey • London • Hong Kong

Published by:

World Scientific Publishing Co Pte Ltd
P O Box 128, Farrer Road Singapore 912805
USA Office: World Scientific Publishing Co., Inc.,
1060 Main St.,River Edge, NJ 07661
UK Office: World Scientific Publishing Co Ltd,
57 Shelton Street London WC2H 9HE

Library of Congress Cataloging-in-Publication Data is available

ISBN: 981-02-3359-0

First Edition

All graphics done in this book are produced by using only a two-button mouse with a 16.7 million digital palette.

Book, Cover and Title Design by Lin Hsin Hsin
Digital images on pp 7, 19, 33, 51, 61, 77, 111, 131 and 167
by Lin Hsin Hsin
Printed in Singapore by Continental Press Pte Ltd

PROPRIETARY STATUS

This book includes some words which are, or are asserted to be, propiertary names or trade marks. Their inclusion does not imply that they are acquired for legal purposes a non-proprietary or general significance, nor is any judgement implied for their legal status.

EXPLANATION OF FONTS

Fonts	**Usage**
Arial	normal text
Arial Bold	poem title
Arial Italic	foreign language, slang special usage of computer terminology
Arial Bold Italic	programs & emphasis

for:

Jek, Jim, Kpp, Tjiok & Yu Gin
who have inspired me
in & out of cybersapce

acknowledgements

'tis a book
written in Real Life
about life cyber

thanks to Jek
from whom I first savor
life cyber

thanks to Jim & Yu Gin
my friends from real & life cyber
who relentlessly *beta*

thanks to kpp & Dr Tijok
my friends from near & afar
who sometimes even tasted the *alpha*

thanks to Bina
who *human checks*
this compilation of techno poems

thank you John Gage, the technological guru
who has so kindly took my manuscripts
as his inflight companion:
from Bangkok to Chiang Rai
then to Zurich, across the Atlantic
to Boston, Washington & back to Silicon Valley

thanks to Sun Microsystems Singapore Pte Ltd
the poetic art sponsor in town
for making this book possible

thanks to Shell Companies in Singapore
the established art sponsor
who once again support

thanks to Cisco Systems (Malaysia) Sdn Bhd
for their support
in this technological poetic title

thanks to my publisher
who once again adds this title
to their scientifc repertoire

last but not least
thanks to all my fans & readers
who besides sampling, will continue to support my titles

FOREWORD

I often write as I listen. As I read, I write, not taking notes, but rather writing down my thoughts as if I were having a conversation with a speaker. I'm no longer an audience, but a partner in an exploration. New ideas emerge as the speaker's words become the starting point for imagining and extending.

Hsin Hsin's poetry takes the language of the designers of cyberspace as the starting point for imagining. She does this in ways that I find provocative, amusing, and true. She helps the builders of cyberspace see what they are creating, and by seeing, they change what they are creating.

I know researchers who take their files, notes and writings and dump them on a desk, mixing and shuffling them, picking them up at random. This spontaneous rearranging creates new flights of imagination, new connections suggest themselves, new ways of thinking and new ways of describing relationships emerge from unanticipated meetings of pages and paragraphs.

Reading Hsin Hsin's poetry is like having unexpected conversations in unanticipated meetings with unpredictable people. She has a wonderful ear for the language of the virtual cybercafe, and helps us visit every table.

Poetry is compressed thought, language crafted to make each word count. Each syllable and phrase affects each other, each element of language is placed and balanced with each other.

Hsin Hsin chooses as her base poetic language the vocabulary of cyberspace. Her basis of her evocative work is the technical vocabulary engineers design not to be evocative. She takes the language of dry precision as an instrument of suggestion. She transforms the careful words to create intricate mechanisms of networks and computers into carefree explorers. She breaks down the walls that separate technical vocabularies She plays with words that take themselves seriously.

And she does this carefully, without misusing or misunderstanding the role these words were designed to play. In cyberspace, everything is built upon manipulation of precisely defined symbols. Computer languages are streams of symbols that create machines. These machines are combined with others in an ever-more-complicated hierarchy of devices. To understand these systems, the technology community invents names and descriptions, but the systems surpass the description.

Only by using these systems, can humans begin to understand what they have built. The descriptions are necessary, but not sufficient, to predict how the systems will behave. Language cannot be separated from behavior.

These systems seem to need an intuitive way of understanding them. This intuition comes from working with, and conversing with these machines which know no restraint of physical law. The language that describes the machines cannot be separated from the behavior of the system.

To understand these systems, to develop intuition, we need poetry. We need more powerful languages, and we need new languages. Engineers can learn from poets to be better engineers. Hsin Hsin's poetry helps us as poets help us to converse with ourselves. She gives us startling juxtaposition, unexpected contrast, and new usage, all in the new language of our vibrant cybercafe.

This book is an exciting merger of engineering and poetic creation.

Mr John Gage
Chief Scientist, Sun Microsystems, Mountain View, U.S.A.
November 8, 1997

PROLOG

this is a book
written by a techie for techies

with *Internet,* the focus, the story
basically, scripting this *boring* glossary
is like translating technologies into poetry
though fun unlimited, but at times I am faced with difficulties

terminology, plenty
to humor these hard-core techies
thru' the pages of my surfing diaries
can sometime sound like *Winsock voluntary*

to those who currently
reside on *domain university*
or are out there practicing *hacking* artistry
truly, I'm a nobody

it's a record, a journey
of my experiences real or funny
in understanding the Internet identity
& grasping the future of technologies

when you discover & arrest stuff buggy
debug & *email* me immediately
give me your comments honestly
for I like to hear your opinions frankly

if you need to check my identity
you can locate me on the *Web,* almost instantly
I reside there virtually
no, I mean seriously

if you can't get me. don't get angry
Ping me or my server fervently
for this is the only way, *effortlessly*
to locate those with cyber identity

Yours virtually
Hsin Hsin

TABLE OF CONTENTS

Net.Net

Net Art

World Wide Web

NetFuture

ABOUT THE BOOK

Jul 22, 95

this is a book
written by a techie
for techies

words uncanny & thoughts zany
are not composed for *power up* money
they're *logged in* to tease

it's just to give you a flavor
of experiences I've savored
whether favorable or unfavorable

anecdotes unforgettable
mistakes forgivable
are all *HTML* enabled

this is a book
written by a creature non-cyber
for all buddies cyber

hence words gathered
designed & wrapped under softcover
to be placed on a coffee table

when cyber feelings uncovered
use *HotJava* as your preferred browser
this is yet another site to be discovered

surf whenever
use it as your *homepage* wherever
savor all that's available

this techie, *cyberly yours* forever
would like to meet you
outside this space cyber

over coffee, *HotJava* the flavor
decaffeinated, preferably
hence words will again flow quickly!

in BYTES we *tr@$_v$@l*

Sep 23, 97

in bits we marvel
in bytes we travel

these are the five chapters I charter

Net life
filled with fancy characters
from cyber *genesonics*
to the loud speakers & silencers
to the would-be *kindwunder*
in **e-mail** we utter
while observing ***e-commerce***
& ***Net fraud*** that rumbles
Net crimes that murder

in ***Net Art***
we marvel or suffer
as a contributor or distributor
& find out the networth
of the Net culture

in ***Net.Net*** the chapter
we thank the technology gurus
& the Network administrators
in their attempts to "*fence off*" the hackers

in the ***World Wide Web***
we pause & navigate together
others crawl & disseminate without orders
as we look further

....

we wonder
what is the ***NetFuture***
what next to offer

whatever,
as we nurture & mature in space cyber
these are the bytes that I've sampled
I regard these experiences, invaluable
a non-experience would be a cyber *withdrawal*

I sincerely hope
that you'll navigate through these chapters
both the hardcopy & in space cyber
so to experience this poetry recital:

In Bytes we Travel!

genesonics = genius

geeks & nerds

```
matrix operator * (matrix)
boolean comformable (matrix);
matrix : : matrix (int rows, int cols)
{
 int i j;
 for  (i = 0; i  < num_rows; i==)
    for (j = 0;   j <  num_cols;  j++)
entry [i] [j] = 0.0;
}
```

```
type def int boolean;
{
 class matrix;
 private:
```

```
matrix :: ~ matrix ( )
```

caffeine

Jul 13, 95

nerds & caffeine
are born as twins
with codes in between

when your energy level is lean
just fire up your cgi-bin
it boosts up your adrenaline

caffeine
a tonic better than gin
it sharpens your RAM like a fin

geeks & caffeine
are next of kin
with codes in between

when your creative cells are thin
it pricks your nerve like a pin
it makes your codes shine like mint

to be drowned in caffeine
a whim or a sin?
a way to lose or win?

NERDS

Jul 21, 95

these are nerds, caffeine thirst
live by Mountain Dew second & Cola first
24 hours a day & 7 days a week in space cyber

these are nerds, nerds that thirst
for all things cyber
from *text hyper* to *sites mirrored*
from *icons* minor to *JPEGs & GIFs* that curve
in *pixels resampled* & *RGB colored*
in *resolutions 96 dpi* or lower

these are nerds with nerve cyber
http versed
like poems, non-stop they murmur

these are nerds that surf & turf
ftp to as many servers
download everything they love

with IQ above the Bell Curve
these are nerds that *telnet* by birth
in *Unix* they berth

these are nerds that *cybergloved*
if you *smail,* they clobber
email, the one way to decipher & confer

……………………………………………………………………………………

these are nerds immersed
in space cyber
once they emerge
they feel homeless
life without a purpose
directions without a compass

these are nerds that prefer
to be residents of the *WWW* by birth
cyber relations they shape & kindle
in *cgi-bin* they whisper
with *home page* as identity preferred
by *HTML* they prosper

these are nerds, nerds that thirst
for all metaphors cyber
from birth to love
in health & in sickness
for richer or poorer
till they return to *dust cyber!*

GEEKS

Oct 16, 97

they solve problems or generate quick fix
their ideas continue to click
while their underwear & socks
pile up in a heap

as their fingers continuously jog
to generate software by the gigs
their lenses may become really thick
while their tooth brushes side step & side kick

nevertheless, toothpicks
can always be found amongst *paperclips*

as bad breathe is the result of nights without sleep
it's good that they only e-mail & don't speak

for them, a T-shirt is regarded as the best fit
while their files & trousers are tightly zipped

a crew cut or a punky hairstyle
makes them look sleek & neat

who are they?
they're geeks

without them, technology will not exists
let alone shine & beep!

paperclips are used to reset systems

THE WAY THEY EAT

Oct 4, 96

pizza, three days' old
beritto, tacho & nacho
all are geek food as we're told

hot food becomes cold
cold food becomes hot
all are left untouched in the bow

coke or mountain dew
coffee or cappaccino
caffeine steadily flows

if your friend has not eaten
for the last 24 hours
don't panic, dont disrupt

you can carefully slip & slide through
underneath your door hole
where thin cheese slices can unfold

alternatively,
you can carefully push through
a Nurenburg wurst through his/her door key hole

be sure to remember,
don't shout or knock @ the door
otherwise watch a tantrum unfold!

NETOHOLICS

Sep 14, 95

every morning they wake up
with coffee in their cup
in the *Net*, they surf & gulp

in waves they rub
admist the surfers' club
while searching for all things *pub*

lunch without *interrupts*
accompanied by *kuay-teow* with chilli & *ketchup*
ice kachung with syrup

meanwhile, *vmail* they ignore & snub
fmail they destroy in parts
smail they erase & rub

coffee continuously they gulp
in waves again they're snug
surfing mileage they certainly chart

they live their lives in *routers' hub*
over coffee, the cup
going to toilet, the undesirable & unavoidable interrupt!

kuay-teow = a kind of flat noodle, Fuxian
ice kachung = a dessert with various sweet ingredients wrapped in ice, Malay

NETSCAPE

Sep 5, 95

this is a net
you just can't escape

this is a scape
that keeps information up to date

do not hesitate
enter & navigate

it's state of the art
& the art of the state

this the scape, once you date
you'll be well-shaped

this is a scape
once you enter, you'll be well-informed

it's your intellectual estate
as well as your playmate

beyond these dates
it's your daily companion & your soul mate

this is a net you navigate
page by page
it's where you forget
your nationality, your hierarchy & your age!

TAKING A BYTE OUT OF THE APPLE

Feb 9, 96

touted as the diesel of the Apple
like his predecessor, Spindler
as yet to turn the table

with 69 million dollars losses in the 4th quarter
Sun the Internet power, will offer
a 23-dollar per share merger

but Amelio, the new wrestler
has turned down the offer
instead of, X 1,300 of its 145,000 workers

since Windows 95, the look-n-feel copier
Apple's share has since dropped 7.8%
will the soil in Cupertino
reach a higher scale Richter?
or will it become stable
over a cup of c a p p a c c i n o ?

will the change of ruler
change the fundamentals
of this nerd naturing cradle?

will this ruler become the savior
rescue this sinking vessel?
we shall watch further!

SUCH IS TACT-NO-LOGY!

Sep 19, 97

even dos was purchased @ a price
like a grain of rice

not surprised?

open your windows in the year ninety five
the great business plan contrived

strive on, strive!

exploring the net in one plus ninety five
navi-gates at a price is an outcry!

keeping trying, try!

bite an apple to continue, is the latest drive
indeed, give a byte to apple to survive!

motivation derived!

@mail

我

U

Wahhhh! got such thing hun?

Received: from NTBJ4K2

147.175.111.18

by bestgate.thisway.com (8.8.5/8.8.5)

with SMTP id UAA18465

Content-Type: text/fancy; charset=u

can C everythng eh?

LIFE ON THE NET

Sep 14, 95

life on the *Net*
has its own *netiquette*
use UPPER & lower case accordingly
stuff we're used to traditionally

if you use UPPER case excessively
it resembles shouting, undoubtedly
you can use it to emphasize a single word effectively
it's always nice to be polite and behave gently

if you use all lower case
is resembles an e.e. cummings forgery
after a period. Begin with a UPPER case assuredly
this we've been taught when kids, really

when replying to email
don't repeat the entire paragraph, nanny!
response only to those you want beside the *cr (*>) necessarily
think of hard disk space, have mercy!

think before you talk, talk wisely
talking in this space, is just like talking face-to-face daily
to be part of this social gathering virtually
courtesy, the first golden rule, *newbies*

your signature, sign intelligently
not 7 pages, to show off or being silly intentionally!
leave your contact or quote a proverb, your identity that is,
anything more than 4 lines, the meaning will be lost, honestly

the X-SENDER

Sep 15, 95

from Australia to Alaska
from Austria to Zambia
these are names with 200 @s attached
very simply, emails that tax

though intention novel
information & importance considered
all servers & routers
all recipients & readers suffer

why can't the sender
understand the agony of the *receiver*
& at least *packet* the @s, in number smaller
by country or whatever
so that it's more manageable
to all parties & all users

so that the recipient
would not again be attached
on someone else's *Listserv* or *X-Senders*

SAY CHEESE, SMILEYS

Sep 14, 95

use **:-)** to denote moments happy
or **:-(** to say you're angry
or **;-)** when you need to wink discreetly
to enhance your text *ascii*
to express your feelings generously

these non-wire frames
are little graphical renderings
that speak, seemingly
though guarantee bytes economically
only if you tilt your head to the left correctly

emoticons silly
are weaved to the *Web* & *e-mail* accordingly
though they can be used to retard *flame* ironically
albeit, when overused badly
can also inflame unnecessarily

VOICES AUTO

Sep 15, 95

be it tenor or alto
these are email voices auto
they answer any questions *prompto*

be it praise or debug stories told
complaints or sorrows
email auto replies, contents monotone

be it platform *Unix*, *Mac OS* or *Windows*
whichever *mailto* you follow
reply identical, yesterday, today or tomorrow

same replies to all different fellows
what annoys is a 12-page advertisement "*borrowed*"
mimicking *auto-courtesy-control*

what really annoys is pretentious voices auto
can just acknowledge with a one-line solo
it's better, though the problem remains with the sender!

Sep 23, 97

when you receive an e-mail
if you don't want to reply:

your silence
is a reply & it multiplies

your pretence
is a reply & it's sly

your **indirect** pretence
is a reply to demonstrate your **indirect** ignorance

however, beware of e-mail that *flies*
flies to the real recipient in the cybersky!

your undesirable reply
is at least an honest reply

from nine to five, five to nine
e-mail, excuses & lies fly
though you can try, when you can't get by
your name will be down or up in the cybersky?

indirect *pretence = replying to the real recipient (A) with the entire contents written for an unrelated recipent (B). The sole intention is to insult A with B's "abiility".*

CU2 Sep 9, 95

these are alligators cyber
who email each other
on matters small or big
it really doesn't matter

hey or *hi,* the common way of addressing each other
regardless of hierarchy, rank, file & order
nationality, religion or culture
truly, this is addressing made simple

headers, body text & trailers
now take a different approach & protocol
nevertheless, well comprehended by all
ye users & hard-core

words, English or American
are replaced by numerals
and/or single upper or lower case characters
wherever possible

as an example
2 & *4* are chosen numerals
2 to replace *two*, *to*, *too* and *tool*
4 to replace *four*, *for* and *fall*

characters are adapted also
C & *U* are typically chosen characters
C to replace *see*
U to replace *you*
a sentence *see you too*

....

is thus replaced by *CU2*
this, save time & disk space too
followers quickly wooed!

EMALE

Feb 1, 96

e-mail me
says Daze
on her home page & cave

be brave
so says this chick
who is male crazy

just point & click
drag & drop
plug & play

cheese & smile
hi & greet
e-m e e t & e-kiss

tease & play
hide & seek
without a hiss

just point & click
cut & paste
whatever your taste

say please & pray
unhide & speak
without a hitch

e-mail me
anyone, anywhere, anytime, please
says this pleading chick

just in case...
e-mail me
& make my day

over the Web, you can't be prey
make her sun shine by saying hi! or hey!
& you too can have a nice day!

don't leave home without e-mailing me
have mercy upon
this chick that is eee...male crazy!

VOICES THRU' THE TUNNEL

Sep 9, 95

two online humans
begin with voices seemingly *solemn*
over emails frequent
developing feelings human

when writing a message
hmmmms..., aaahhhs, woos & ehs
yip, yeap, yeah, yap, yup, yupe
nope, noop & noope...
are all nuance musts
piping thru' the mail passage
to carry a voice
to understand signals

it's only natural, feeling internal
expressed thru' this cyber tunnel,
really, this is only a concept fundamental
resident in human kernel

feeling, the master
happy or angry
sad or frustrated
truly, doesn't go away in space cyber!

BACK TO BASICS

Sep 9, 95

two on-line beings
one male, one female
who meet over email
can develop a relationship, male-female

two on-line beings
over a prolonged period of being
can develop a feeling
otherwise vacant or missing

cyber beings
are actually human beings
with blood, flesh & soul
indeed, a total being

when seeing without touch
& talking without saying *ouch*
can only last for the time being
stretching it too long can be overbearing

bond phantom must come into being
mailer-daemon becomes itineraries flying
routers re-route
beings ascend into clouds

when routers repeatedly re-route
voices echo, spurt & sprout
this is when nuptial knots tie & love shouts
three, indeed is a crowd!

cyberspace the rainbow behind the cloud!

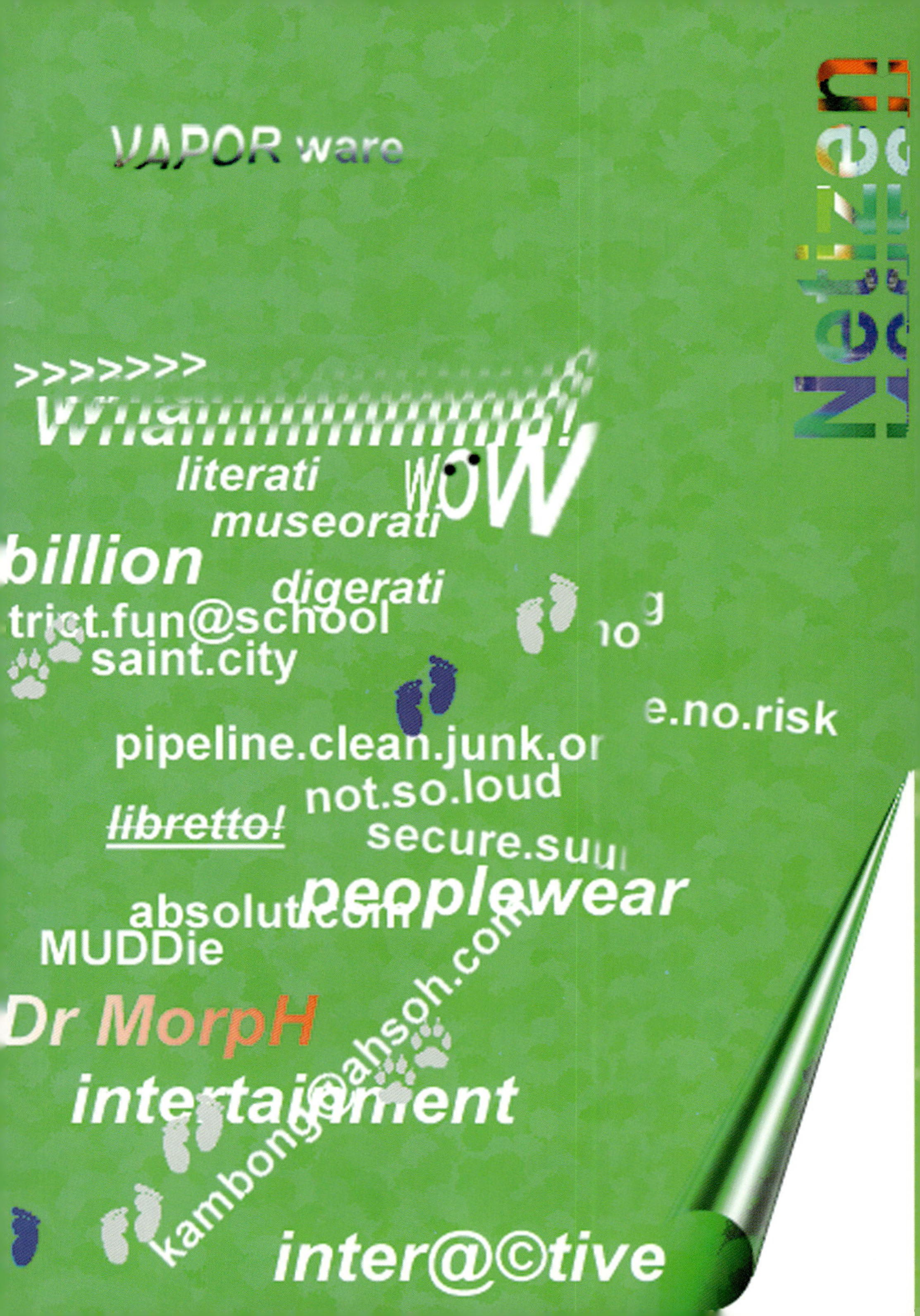
Netizen
VAPOR ware
>>>>>>>>
literati
wOW
museorati
billion
digerati
trict.fun@school
saint.city
e.no.risk
pipeline.clean.junk.or
not.so.loud
libretto!
peoplewear
absolut.com
MUDDie
Dr MorpH
intertainment
kambong@ahsoh.com
inter@©tive

real life presents cyber life
cyber life represents real life

inter@©tive

Sep 18, 97

exit your real life
get a life
on the Net
have a chat
wear a different hat
enter into a new lease of life
& that's life!

MUD & MOO
woo & coo
it's cooooOLLLLLLL !

fool & to be fooled
sue & get sued
flame & spam
lies & truth
servers push, clients pull
choose & to be chosen

stay tuned & get frozen!

CLICK HERE

Sep 14, 95

click here
if you want to know more about me
or see my picture, dear

so, click right here
see no picture
click here again, dear

so click again
sometimes it returns to first *click here*
at other times, a *404 URL not found* smears
don't mind even if a picture of a dog appears
in order to continue with another fear
of the **awesome** *click here*

is this awesome
or is this funny?
read here, dear

behind the good Web pages
there stand these clicks
or rather tricks

time wasted
interest lost
patience drains, dear!

USER INTERFACE

Jul 29, 95

geek & machine
have nothing in between
at most, separated by a screen

geek in T-shirt & blue jeans
when sees & meets a girl *off screen*
his face turns green

geek, social skills lean
with *non-disclosure* in between
does not display emotion off screen

when geek meets Jean
he's keen
but ain't treating her like a queen

when the system hangs
opportunities like this ruined
geek quietly screams

nevertheless, reboot,
geek again turns on his screen
& waits for the next *user-friendly* Jean

HEAVY DUTY

Sep 15, 95

these are pages, heavy
80 Kb, to say the least
though contents rich & with quality

when transmission doesn't perform its duty
users like you & me
suffocate daily

these are pages that emphasize quantity
"*must be comprehensive*" -- the mentality
even if you reside on your own server, there're still worries

these are loonies
who *UL* or *DL* as many
by subject, by source, by links, by matters funny

from one to infinity
as long as appearing on my *Homepage* mandatory
I'm happy & you are merry

considering even a fifth of *80 Kb*
it's a size unhealthy
let alone conform to normalacy

as much as we appreciate your quality
please cut down the quantity
ie, be *http-friendly*

HARK THE BARK!

Oct 4, 96

this is a calculator
that barks or meows
depending on the choice of its owner

it displays numbers, from 0 to 9
it provides operators, for functions numerical,
from plus to divide
decimal point & operator equal
even square root & reciprocal
keypad looks complete & normal

you just need to press your desired numbers
& the required operator(s)
once ready, simply click equal

however,
depending on the bark or meow mode
that you've selected earlier

the return result will be depicted by
the number of *wooh-wooh(s)* or *mmm-ee-ow(s)*
you get from your browser

you'd be amazed
it's properly barked or meowed precisely
so, listen attentively!

NO MORE MESSING AROUND!

Oct 3, 96

these days you can rear
dogs & cats, ducks & chicks
in this cyberspace
indeed, name the pet, take your pick

enter chicken feeds
day by day & bit by bit
you'll slowly see chicken feet
complete with feathers & beak

enter green-worm feeds
day by day & bit by bit
you'll see daisy duck web
complete with wings & waddling feet

if you stop the feed
they'll stop growing, indeed
if you overfeed
they'll definitely grow fat

if you provide an unclean feed
they'll continuously shit
you'll then get a feeding guide sheet
& be told to clean & wash their feet

between their daily feeds
you must stroke their feathers or fur
hold their paws or feet
talk to them, indeed
only when you do the correct deed

....

they'll entertain you
with song & dance
using their paws or feet

when you need to go out of town
you can leave them with your most trusted feeds
tell the system when to auto-feed
& your favorite browser will gladly takeover, indeed!

The cyber *kia-sus*

Sep 19, 97

whatever you can do,
we can do better & faster

whatever you can do,
we can redo, undo or outdo

before & soon as there's a new Web or Net feature
we'll make sure we feature

when *blink* was first blinked
we all blinked until you shouted "*stop blinking!*"

when *frame* was first mentioned
we all framed until we heard "*unframe*"

when *shockwave* sent its first wave
we were shocked to see the tides & waves

when *Java* brewed its first froth
we all came forth

when *ActiveX* was activated
we too get activated

when *Flash* gathered its fan
it was more than a flash in a pan

whatever you have at hand
we'll again push the clock's hands!

kia-su, literally means afraid to lose, Singlish

WHATEVER U'VE, WE HAVE IT TOO!

Mar 28, 96

whatever U've
just as in real life, we've have it too:

we exhibit cyber-**a**rt
live on cyber-bread while riding in cyber-**b**us
we accept cyber-**c**redits
& credit U with cyber-**d**ebits
in this frontier without cyber-**e**nd
beware of cyber-**fr**auds
clothe in cyber-**g**ears
& watch out for cyber-**h**ackers
constantly cyber-**i**ndulge
filter out cyber-**j**unks
seal with a cyber-**k**iss
& enjoy all the cyber-**l**ove
fill your inbox with cyber-**m**oney
don't get flamed with cyber-**n**ails
look out for cyber-**o**pportunities
get ready & cyber-**p**ark
be a cyber-king or **q**ueen
take cyber-**r**isks
& bet with cyber-**s**ecurity
make sure U trap all the cyber-**t**hieves
promote cyber-**u**nity
wipe out the cyber-**v**ibes
be prepared for a cyber-**w**ar
screen out the cyber-**X**-rated
well, it's time for a cyber-**y**awn
while breathing in this cyber-**z**oo!

REACHOUT : OUTREACH

Sep 19, 97

on the Net
free is the culture
free-of-charge is the venture
advertisement is the cuture-vulture joint venture

the Net
is beyond an *ad*vertisement *venture*
*it's an ad*ded ven*ue to* ma*ture*
so it's an adventure

on the Net
there's a cult
a *cult* who ven*ture*s
so, it's *culture*

the Net
displaying an advertisement is an avenue
for every click on an *a*dvertisment
so, it's re*venue* for the *venture*

the culture is:
to be free is an adventure
to create a venture is to gain revenues
in & out of cyber avenues

VOLUNTEERS

Sep 19, 97

be it aesthetics or statistics
you can now gather free clicks
by first creating Web pages, hip
the idea is: to get hit & generate traffic

write a script & analyze the submit
design & implement your ideas, big
based on this incoming traffic
frolic the submit & generate new outgoing traffic

free mileage, high traffic
project fund seek
display your innovation in public
get applause & gather more new clicks

old visitors retire quick
but new visitors will arrive quick
when visitors arrive by the heap
you can justify to buy newer & better chips

while clicks are regarded as carrots
form-filling & e-mailing do not entail whip
Net kids are easily found & they come cheap
a great variety & without warning beeps!

REAL vs CYBER

Sep 23, 97

what's the difference
between the real & the cyber world?

both are proper
both are improper

both are in order
both are out of order

both give orders
both receive orders

both are logical
both are illogical

both are fighters & warriors
both are commandos & soldiers

both disgruntle
both wander

both wonder
both do wonders!

NOISE

Sep 23, 97

if you have nothing new to offer
don't re-iterate, don't mirror, don't bother
in this space cyber

if you have nothing new to offer
don't copy, don't rumble, don't fumble
in this already overwhelming place beyond the earth

if you have nothing new to offer
don't tamper, don't loose temper
in this already noisy chamber

if you have nothing new to offer the world with textures & fibers
don't export yourself to the world cyber
in this already congested space ether

always remember
filter before you offer
so that others won't suffer!

redit card # {[(&*()%^:"_x,_ß+_dÆ)]}~
can unscramble?
a piece of cake!
talk softer!
e-commerce
U can shop 4 virtually anything!

THEY FOUND THEIR WAY?

Oct 6, 96

be it service, product or display
so long as they can grab your e-mail
they will find their way into your e-trays

everyday,
when you open your e-in-tray
you'll find some xtras on the way

with 381K Web Sites & 54 millon Web pages
it's better they pronounce product excellence
with their URLs prominently displayed
in your e-in-tray
otherwise, they may just get decayed

it began with an advertisement in the s-mail
then it became f-mail
& has since progressed to e-mail

advertisements, the **20-page** e-mail way
is a way to increase traffic volume
along the information highway
so that you can drag them into your e-trash tray
& recycle the polluted electrons on the way

TAKE A WALK THRU' THE MALL CYBER

Sep 9, 95

from fresh flowers to towels
from software to porn cyber
these are shops on the mall cyber
open for business 24 hours

catalogs, from cover to cover
cyber gloss samples
just click, it's instantly available
to all shoppers cyber

ordering is made simple
credit card number you enter
acknowledgement appears instantly
on your RGB monitor

is security something that matters?
depends on their integrity,
& how you value your credit card number
& the hackers you may encounter

goods you ordered
will they arrive in order?
depending on whether
you meet cyber hijackers

will your cash become vapor?
depend upon how the transaction is handled
jokes aside, humor honored
basically, trust is the only crux of the matter

electronic commerce
will or will not florish depending upon
how fast security hubs & servers
can be turned into vaults cyber!

GAMBLING IS LEGAL

Oct 3, 96

TOTO or horse racing
Russian roulette or Black Jack
solitaire, mahjong or one-arm-bandit
all can be played in this space cyber

be it Explorer or Navigator, get on to your browser
enter your name & personal ID proper
no pen & paper, betting is between you
& your downloadables

after you logon, you will then be greeted
with a croupier cyber
wearing a smile wonder
with a skirt leather & even shorter
whatever the length, *presto*
it is completely adjustable!

now enter the amount to bet on, gambler
followed by your mandatory credit card number
or your bank account number
when you're ready, simply press enter

dice in the tumbler
or horses on the e-tracks with e-equestrains
it comes complete with noise & gallops normal
volume is completely adjustable

pull down your jackpot handler
hit seven e-lemons, e-coins will flow in order
hit seven different e-fruits, don't bother

....

bet on, your e-luck might just be better
don't forget, the game is open 24 hours
unless something goes wrong with your server!

these days
you can e-bet from Macau to Morocco
gambling is available in cyberspace global
without flying to Genting or Vegas in Nevada

however, the inheritance [properties] of gambling
& the behavior of the gambler
can't be changed ever --
you can still fumble & topple
& be caught with your pants down,
in the middle of this space cyber!

E-MONEY

Oct 3, 96

when money turns itself into bits & bytes
it travels around via satellites
fiber optics & the like

stored value & rechargeable smartcards
digital signature encoded smartcards
are *e-money* & *e-trusts*

transaction will be done over the Net
via credit cards & smartcards
in your *e-wallets* & *e-safe-vaults*

your e-cash & assets' information integrity
& their circulation security
should be assured in your armchair tranquillity

guarding the clients' information assessibilities
would be the utmost priority
for any *e-financial* authority

this is when you'll see
e-bank erases the concepts of
merchant, investment & off-shore banks

this is when you'll see
e-smartcards e-merge with credit cards
& you'll truly walk around without money

....

nevertheless,
you'd beware of *e-robberies*
e-frauds & *e-conspiracies*

the day will come
when your servers & hard disks
will be confronted with fearful e-risks!

VIRTUAL OFFICES

Sep 9, 95

these are offices
once housed in buildings tall
since Internet, responded to cyber calls

they still occupy real mode floor
with real or fake pictures on the wall
but footage has become small

there are still people
traveling up & down the corridors
elevators, escalators and all

there are still phone calls
but vox mail answers them all
while snail mail continues to crawl

there are still people hanging around outdoor
outdoor? consider as an official call
a luxury, a concern to fellow being above all

commuting, now being considered a chore
travel only when leisure calls
"*yes, boss*", only when his email calls!

Net Crimes

NO MAN's LAND

Sep 14, 95

copying a movie
giving it to a friend
or putting it on *LAN*
is it an offence?

save it on a local hard disk or server, *c'est le même*
is it avoiding or rescuing a traffic jam?
post it as news, put it on the *newsbench*
is it a disguise or following a trend?

intellectual property in cyberland
whether with or without a brand
liberates itself & descends
to *knowledge-&-curiosities-quenched*

old age copyright & trademark laws can't bent
themselves to new age technologies to defend
prank or genuine interests and those with good intent
it simply occupies no space in this cyberland

while archaic laws remain
lawyers must catch up & disciplines be retrained
to be useful in this *Internet* domain
but when?

PAPARAZZI vs PROXIES

Sep 16, 97

What's the difference between paparazzi & proxies?
both deal with issues of public & privacy

while paparazzi aim at an individual's privacy
proxies aim at the masses' privacy

paparazzi pry
proxies defy

while paparazzi aim at celebrities
proxies aim at the ordinary

paparazzi depend on telephoto-lens availability
proxies depend on digi-lens availability

while paparazzi exhibit media vulnerability
proxies inhibit private activities

both produce documentaries
both write history

while paparazzi upload personalities
proxies download activities

both press dairies
both suppress dairies

....

what's the difference between paparazzi & proxies?
both are rudimentary to public enemies?

paparazzi *n.pl It. The word "paparazzi" was coined from the word "paparazzo" between 1965 to 1970. "Paparazzo" was the name of the photographer in Federicio Fellini's movie "la dolce vita" produced in 1959.*

TO SCAN

Jul 6, 97

there's a device named a scanner
justifiably used to scan to archive
scan to correct & modify

photographs, transparencies & slides
all can be scanned,
publish it to the Web to glorify

others' photographs, transparencies & slides
all can be scanned,
to edit, correct & beautify
to distort & terrify
to cut & paste, resize & call it a new design
to morph, tween & call it thine

scanning others' images
is only an exercise
hoping that no one will ever recognize
the originator & original published design

scanning well-known images & design
is a sure way to scoop a gold mine
it's fast & no one will ever mind
who is the originator & what the original design was

with digital watermarks & fingerprints
scanning copyrighted images & design
is a sure way to get digitally cuffed hands realized
culprit's face scanned & on the Web infamously publicized!

cloning?

Mar 7, 97

while sheeps can be cloned
you can leave digital art cloning alone

be it triangle or cone
fine arts on the Web
are created by fingers with skin & bone
they're not to be copied over a modem-phone

Jane or Joan
once caught, access privileges will be sewn

in the name of research

Sep 21, 96

in the name of research
they crawl & search
to fulfill their urge

Web lifters, once they're caught, excuses sought:
"*is you who didn't close your Web doors or ports*"
so they scholarly thought

when your server is hoarded by their crawlers
because you don't know & did not exclude the "*robot.txt*"
hence they bear no faults
& you need to be taught!
So, don't complain & don't be distraught
as "*to teach*" is their job & university is their resort

when they're confronted:
your domain name and country code they sought:
"*we can't be learning from you?*" confidently they thought

when they're asked to surrender:
they argued that they only copied text
& they regard this as their permissible default

in the event your lawyers need to see them in court
the war needs to be fought
for you can't be the innovator, "*full-stop*"

as my daily checks on the access & error logs
tell me that cyberlaws are as dead as logs
sad to say, old laws can't be learned by old dogs

nevertheless **CPYRGHT101** still need to be taught
to the professors that have been caught
or those potentially possibly be caught

Web lifters' abuse and bad habits need to be taught
they must prohibit, they must guide & lock
Web lifters need to learn: once they're caught
copyright acts are no longer lawful exhibits,
but awful digital handcuffs and cyber jail-locks
they can put you in a real life dock!

to prohibit, there is no other ways:
you either need to inhibit your Web exhibits
and/or to invent & install your own Weblocks!

outell = out to tell

WEB LIFTERS

Sep 21, 96

these are Web trotters
that travel via crawlers
to check out who has the best offer

once they identify the model
they become the professional Web lifter
rather, professor of Web lifters

after they've uplifted the entire Web Site model
they store it in their refrigerators
meanwhile, only they get to see the copied model

thus they examine the model
from all sides and all angles
to find out the secret recipes of this model

after they get out of the deep-freeze
they select whichever angle they prefer
& showcase in their own library digital

they can then write their Ph D thesis based on this model
publish their research results in reputable journals
& make their name be known as the Internet leader

these acclaimed human-computer-interface scholars
indeed, seasoned Web lifters
are professional Web looters

from the Perl-scripted Webcopy to the meta-crawler
nothing is works original
thanks to the cut & paste, warp & morph tools available

however, when this is seen as works from scholars
it becomes a different matter
it is an innovation, says the admirer!

DEBIT OR CREDIT?!

Sep 19, 97

make a hypothesis
on the Web, gather free aesthetics & statistics
make an analysis
write a Ph D thesis

on the basis
of developing a new search engine (un)necessary
to benefit everybody
to commercialize it & make money

is this the virtue of the Net
or the best way to conduct research & analysis?
to kiss your intelligence
& downgrade others' diligence?

is this the best way to develop
the Net of the future or future of the Net?

This will depend on
the calculated risks of the candidate
the cyberlaw in the making or breaking
& Netizens' approval or disapproval

but before long
a Ph D thesis without reference credits
will debit
the forgotten source of aesthetics & statistics!

TOGETHER WE TRAVEL & CHECKMATE?! Sep 18, 97

this is a place
where *digi-fingerprints* & crawlers mate
make a date
digi-fingerprints & crawlers become travelmates

both are innate
both are *robomate*

while one copulates
the other regulates

one copymate
the other writes the fate

while one loves
the other hates

together in cyber court
they debate

stalemate or checkmate?

PUNISHMENT -- UNEARTHLY

Oct 3, 96

if you commit a petty theft, fraud or crime
in this space cyber
you need to be punished
according to the cyber law & order

soon as you're proven guilty
you need to surrender your personal particulars
depending on the extent of your offence
you can get an e-fine, digital handcuffed or e-jailed

to pay an e-fine
you enter your bank account number(s)
&/or your credit card number(s)
together with the respective expiration date(s)
your fine will be debited accordingly
if any of these fields is missing or entered incorrectly
this Web page will be activated automatically
whenever you open your browser, daily

when you're digitally handcuffed
your IP address(es) will be displayed prominently
on all bulletin boards
wherever & whenever there's an opportunity

your real life particulars will be transmitted
to our lawyer, injunctions & hearings
will be carried out accordingly
to the on- & off-line communities globally

whatever the case, you'll be e-jailed definitely
depending on the severity of your offence
you won't be able to logon for x-days
no more surfing no more e-mail, unfortunately!

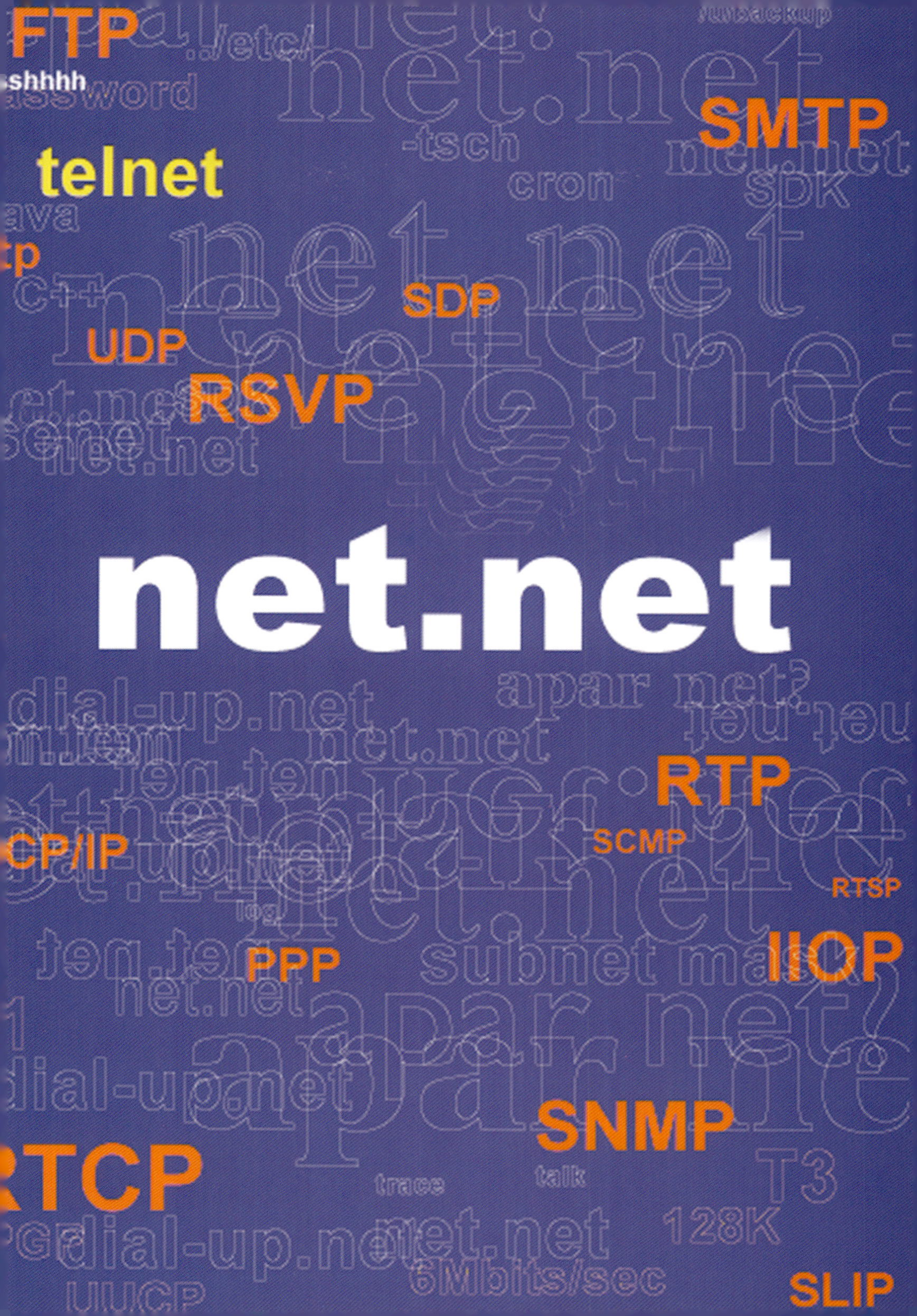
FTP
shhhh
telnet
SMTP
SDK
cron
SDP
UDP
RSVP
net.net
dial-up.net
apar net?
RTP
SCMP
RTSP
IIOP
PPP
subnet mask
SNMP
RTCP
T3
trace
talk
128K
6Mbits/sec
SLIP
UUCP

without the Net
we'll be trapped in the net

The Net

Sep 18, 97

The Net is a medium
that implicates

it
duplicates
triplicates
...
n-plicates

as it replicates
it complicates?

as we innovate
& renovate

we indicate
LIFE on the Net!

as a result

it intoxicates
& we suffocate?

expliquez si vous plait?

SOMEWHERE

Sep 20, 97

the Internet was born somewhere
sprouted & spawned everywhere
it continues to be nowhere

beware of its frauds, virus & *insaneware*
by guarding its *underware*
using the most current software & *faster-than-ever* obsolete hardware

be sure you own a home(*page*) somewhere
so that the world is aware
of your *intelliware* or *creativeware*

while transacting your favorite hotels
or fluffing your dog's hair
make sure your credit card # is still in the *vaultware*

so that you can call this vacuum
your **home** somewhere
guard it fervently to prevent *hackerware* & tear

at nite, you'll then able to sleep soundly
hugging onto your teddy bear
while packets tunnel through your *bastionware*

RUSH OR FLUSH?

Nov 15, 94

be it voice or text
audio or video
images virtual or real
format single or mixed

these are bits and clips
that we draft & craft
in and out of the office & studio
pantry & patio

these are bits and clips
built on chips & chips
that we submit & transmit
others may omit or remit

these are bits and clips
the capital IT
that we say great & use IT
but others may say damn IT!

from must to lust
this is info we trust
but others may frust

from superslow to ultrafast
this is info we rush
but others may flush

....

from must to lust
this is info we compress or prepress
but others may suppress or get depressed

from bits to clips
from truth to shit
this is info that we dedicate
but others may eradicate

from bits to clips
from truth to shit
this is info that we dictate
but they suffocate

nevertheless
bits & clips, truth & shit
information highway summit
we admit & submit

Comdex/Fall 94, Las Vegas

PASSWORD

July 22, 95

when you subscribe to a service
they screen for an identity
so you need to create a password, absolutely

these are words necessary
to identify eligibility
to gain and safeguard entry

these are words when being asked
it's your task to provide
it's mandatory

these are legalities
that filter, parse & pass
to prevent undesirable encounters

don't use words found in the dictionary
or anything simple
to avoid hackers' artistry

employ password algorithm
create one that's tough to remember
& tough to unscramble

back 'em up in your memory
— the best place to store your keys
for easy retrieveability

but, when you've one password too many
write 'em down when necessary
so when you forget, you're still ready

when you've one password too many
save 'em in the disk when necessary
create a password to gain this entry

when all else fail
just laugh at yourself ridiculously
while you look for hackers suspiciously

seriously, you can't change 'em
when you forget 'em honestly
email system administrators or *websters* nicely
so to recreate password per service
change 'em periodically
so that you can forget 'em readily?

password: a techno-haberdashery or a measure of security?
websters = webmasters, webmistresses

just like in the real world foreign services core
on the Internet
you receive or transmit your requests via ***protocol***

using the telephone lines,
it began with the age-old ***S**erial **L**ine **I**nternet **P**rotocol (SLIP)*
and the now commonly used ***P**oint-to-**P**oint **P**rotocol (PPP)*

machines communicate to one another via
the ***T**ransmission **C**ontrol **P**rotocol (TCP) **I**nternet **P**rotocol (IP)*
not forgetting the dino ***U**nix-to-**U**nix **C**opy **P**rotocol (UUCP)*

to download or upload files, use the ***F**ile **Tr**ansfer **P**rotocol (FTP)*
browse the Web with ***H**yper-**T**ext **T**ransfer **P**rotocol (HTTP)*
or start a new session with ***R**eal-**T**ime **S**treaming **P**rotocol (RTSP)*

to exchange integers or arrays or objects reference call
or to implement *CORBA* solutions over World Wide Web
use ***I**nternet **I**nter-**O**RB **P**rotocol (IIOP)*

if you're a techie hardcore
prefer to log in from remote, of course
simply use the ***Telnet*** protocol

since 1992, the adopted Radio or TV paradigm
allowed the set up of the Internet ***M**ulticast **B**ack**bone*** (*MBone)*
to establish and send *multicast address*

....

tune into a *multicast address*
multicast routers setup *tunnels*
by sending a ***I**nternet **G**roup **M**anagement **P**rotocol (IGMP)*

not forgetting the original
***D**istance **V**ector **M**ulticast **R**outing **P**rotocol (DVMRP)*
that unicasts & generates routing tables

to *unicast or multicast,* packets use ***U**ser **D**atagram **P**rotocol (UDP)*
to map port numbers to processes
sessions are identified by ***S**ession **D**escription **P**rotocol (SDP)*

like the ***I**nternet **M**essage **C**ontrol **P**rotocol (ICMP)*
***S**treamed **M**essage **C**ontrol **P**rotocol (SCMP)*
is used to control message congestion over *node-to-node* flow

identify source & payload type
then deliver packets or encryption key
by ***R**eal-time **T**ransport **P**rotocol (RTP)*

participants and session data
based on participant count & reports frequency
transported via ***R**eal-time **T**ransport **C**ontrol **P**rotocol (RTCP)*

check and signal resource availability & establish policy
to exercise admission control and grant permission control
thus controlling the traffic flow via ***R**e**S**er**V**ation **P**rotocol (RSVP)*

while the IETF is still resolving the outstanding
accounting & charging
authentication & access control policies

there's still the *__S__imple __N__etwork __M__anagement __P__rotocol (SNMP)*
an application-layer protocol
facilitating management information exchanges between network devices
& I need to make an economical call
neither via the *__C__lient-__t__o-__C__lient __P__rotocol (CTCP)*
nor the *__I__nternet __M__essage __A__ccess __P__rotocol (IMAP)*

but just e-mail
using the *__S__imple __M__ail __T__ransfer __P__rotocol (SMTP)*
instead of issuing an IDD call!

multicast address starts with 224.0.0.0 to 239.255.255.255

firewall

Sep 20, 97

a network exposed on the Internet
if not properly protected
will be under attack

you must set up firewalls to protect
against intrusion & destruction
as well as outright break-ins

be it Gopher or Archie,
Telnet or FTP, Finger or Whois
the intruding-experts leaves no trace

determine network administration policy
monitor & build redundancy
like setting up a server proxy

allow e-mail to only trusted machines
disallow to others
besides keeping away & hiding critical servers

log & trap, trace & surface
all connections & break-in attempts
to protect & prevent hackers & misconfigured computers

purchase & install algorithms firewall filters
to drop or deny, pass or permit packets
according to your criteria & requirements

set up firewall authentication
as an authentication server
so that IP can be correctly destined & filtered!

IN YOUR SHELL, WE DWELL

Sep 12, 95

it doesn't matter where you are
so long as you can *telnet* to where they are
you can be one of them, aaahhh...

in their *Unix shell*
you happily dwell
in love you fell
vi, *emacs* or *pico*
in *C* you spell & tell
chmod u+x, show & re-tell

outside the traditional tools used in Bell
you can also enter the *Perl shell*
script in *Perl* & stay in that cell

Perl codes, *pas belle*
but its appetite, big like ..ll
swallows character strings with unlimited length well

when everything jells
store it away in your resource well
disconnect & exit into your own well

THIS IS HOW YOU SEE LIGHT

Sep 9, 95

these are lights
powerful as, if not more than traffic lights
they tune to your appetite
they manage your sights

click & you'll see light
sit tight & watch these seven lights
while you cringe & your fingers bite
they count your bytes

once you power on
you'll see the *CS* light
click on your *TCP/IP icon*
connect & see other lights

once dialed, you'll see *TR* takes its ride
the third being lighted from your right
followed by the *ARQ/Fax* light
& the *CD* light

while the *pulsating comet* strikes
you anxiously wait for the *SD* & *RD* lights
to *handshake* & take their flights
you again cringe & your fingers bite

it's only when the *RD* & *SD* take their strides
you can then look @ or download your favorite bytes
or show off your pride
from whatever Web sites
when either of the *RD* & *SD* lights go on strike

....

you get upset, & *IT* bites
indeed, information in or out of sight
much depends on these mighty lights!

dedicated to David Lim

WHAT'S NEW? WHAT'S HOT? WHAT'S COOL? Sep 9, 95

everyday, you find out
what's new?
what's hot?
& what's cool?
as you crawl in & out
the heaps of Web sites

everyday, there will be more
what's new?
what's hot?
& what's cool?
as you waylay & side track
the avenues & sidewalks of these sites

every other day, someone will take over
what's *new*er?
what's *hot*ter?
& what's *cool*er?
as you park on these sites, you then realize
the older ones are no longer *new, hot* & *cool* sites

every other day, someone else will again surrender
to what's *new*er than *new*er?
what's *hot*ter than *hot*ter?
& what's *cool*er than *cool*er?
and force to be replaced by
the mighty new driver

so everyday, you find out
what's not *new?*
what's not *hot?*
& *what's* not *cool?*
as you crawl in & out
the heaps of Web sites

SEEK & YE SHALL FIND

Sep 10, 95

from *harvest* to *manifest*
from *Einet* to *jumpstations* & all
you query & search results pour

from *Infoseek* to *Nikos*
from *Yahoo* to *Lycos*
these are engines packed with logos

from *http* to *ftp* to *Gopher*
they consider *and*s before *ors*
adjacency in its relevancy above all

you search by keywords, words, phrases
or whatever the cause
at no costs

you define search criteria & minimum score
they search by weight & other scores
7,275,105 unique *URLs* explored

search results they endorse
& you, sit, watch & applaud
while *URLs* appear right at your front door

click to read & explore
return to the search floor
again click & see more

....

search result, absolutely no flaw?
it depends on the method deplored
once in a while, jokes surface & laughter draws

nevertheless, gulp more caffeine
thus, forces you to query again & search encore
indeed, search ye & ask no more!

 # as of Sep 21, 95 Lycos

POINTING AT MY FINGER

Sep 10, 95

if you want to follow my outs & ins
just *finger* me in my *cgi-bin*
instantly, you know when I log in
& what I do in between

I just define my variables
change my mail it handles
my coming-in & going-out matters

it's in this bin
I leave my *finger prints*
for you to know when I was last seen
right in front of your screen

point at my *finger*
see where I linger
find out my mail arrivals
& have I read them at regular intervals?

though you won't know
whether it was 1 hour 23 mins, I was kept idle
or I went into the Web jungle
or whatever tasks I mingled

what you would know
is the address of my logon terminal
my whereabouts, you no longer puzzle
& your IP address, I am no longer baffled!

....

if you miss my voice, real or those on the *recorder*
reach me by phone as you *finger*
save your wrinkles
give me a tinkle!

IT's TIME TO GET OUT

Sep 10, 95

everyday you go in & out
the *World Wide Web* without doubt
you also stick your head in & out of *cyber clouds*

for places far away from the madding crowd
more often than not
you experience *Netscape time-out*

between these 3 to 10 seconds *time-out*
it's best you sip your coffee or step out
to catch a breath or check other stuff out

but more often than not
you quietly shout
"*not again, shit! time-out...*"

though *Trumpet Winsock*, *Chameleon* and *BSD Unix*
adapts, computes retransmission *time-out*
& quickly adjusts itself to a maximum performance out

surely, it can't go on forever adjusting itself at frenzied bouts
it will eventually make a turn about
& inform you, it's time for real *time-out*!

reaching out the far side

Sep 9, 95

these are sites
that lie on the far side
with precious bits & bytes
downloadable throughout the day
& during the nite

you can hitch a free ride
by *ftp*ing to these sites
& get free stuff downloaded to your site
as long as you've been granted permissions to read & write

from *betas* to the *version riped*
just observe the *downloaded rights*
don't resell, else violate the *copyrights*
*copy*ing in this manner is only exercising your *rights*

from *betas* to the *version riped*
the only worry is when *port 21*, or *2021*
1928, 2484, 13134... do not respond
or permission is denied

still, the real injury during your free ride
is when "*unexpected internal config error*" strikes
during transmission of the last 4k bytes
this truly hurts your eyesight!

not unless you use the *checkpoint restart* to rewrite
depending on the filesize, you can only sigh or cry
re-*ftp or* try other sites
just hope for the best, right?!

A-pPROXI-MATELY SERVED

Oct 6, 96

fetch a page via your Web browser
you'll be in touch with the host server
your desired page should be in your disk local

however, some authorities
set their disapprovals
for the very reason, simple

Web pages & Web Sites
they don't favor for reasons moral, for example
you too, shouldn't savor

thus any Web downloadables
must first be transfered to the proxy servers
@ your Information Services Provider
before it reaches your *cache* local

any IP addresses with pornographic flavor
will be censored
& suspend the offensive surfers

with certain technological power
it's easy to disguise & hard to distill such flavors
in this fast growing cyberspace without border

even with these proxy servers,
there'll always be many *"runners"*
& the already slow bandwidth will become even slower

....

& for those who must read some such numbers
like the Bill Gates' interview on the playboy server,
please read them from other servers

& for those concerned authorities who need to deter
such offensive flavors
the only way is to make these surfers responsbile!

ROBOTS & CRAWLERS -- FRIENDS OR FOES? Oct 16, 97

there are robots
Spiders & Wanderers
Worms & Brokers
Bots & Web Crawlers
Web Ants & MetaCrawlers

they crawl to temporarily reside
or rapid-fire to excite
building Web wide indexing
on millions of documents
on millions of sites

whether you are up and awake
or in deep sleep or *snooOring...*
they're constantly wandering in your site
though they're physically out of sight
they're digitally trailed on site!

if they *rapid fire* or bash
genuine visitors of the Web site
can get "*refused connections*" from the site
with "*high load*" for the server owners side by side
hence slow performance or even a system crash

you can exclude these robots or Crawlers
by disallowing these *User-agents*
or include the latest *tag*
<META NAME="ROBOTS" CONTENT="NOINDEX">
in your favorite URL source text
to deflect & reflect!

PRETTY GOOD PRIVACY

Sep 14, 95

if you worry
you need peace
while you talk in confidence & tease
punctuate with humor & please

don't take chances, never risk
digital signatures, the best strategy
subscribe to *PGP*
to ensure pretty good privacy

you sign with your *secret key*
encrypt with his *public key*
he decrypts with his *secret key*
& authenticates your signature with your *public key*

use cryptography
for your data security
to ensure data integrity
the way to enjoy reading in plaintext 8-bit ascii

PING PONG

Sep 14, 95

set your preferences
iteration number, whether discrete or continuous
data length, 64 bytes the normal
elapsed time between transmit & receive intervals

if you *ping*
it pongs

ICMP packets echo
bytes sent & *bytes received*
at your preset interval
calculates round trip min, average & max ms

you can start or stop
this *Ping* song to whichever *IP address* or *server*
as long as you don't hear *packet loss*
at regular interval

packet loss,
strikes like a gong!

VIVA-LA JAVA!

Sep 20, 97

originally, the intent was to use C++
but creating a new language is an even a bigger plus
hence Java

the platform-independent interpreted language
& an operating system that multithreads, multitasks
& runs fast

hence Java tools are created by Java
& Java-enabled browser, HotJava
so that Web pages can gyrate & emit lava

have a *supper-cuppa* Java
the object-oriented, distributed, robust & secure
the architecture-neutral, portable manna

Java,
the compiler, interpreter & debugger
disassembler & header file generator
JavaDoc, Applet Viewer & Java-enabled browser
viva-la Java!

JAVASCRIPT

Sep 20, 97

this is a Script, Java like
it can be used on the client or server alike

specify the language attribute
<script language= "JavaScript">

data types & variables
integers & literals

Booleans & floating points
strings & Escape Characters
expressions & operators
just script them together

it can create instant wonders
on the Web it travels

use *Event Handlers*
onClick or *onMouseOver*
onSelect or *onSubmit*
all can be triggered

write *generic cookies*
create a Menu Manager
use address IP
to generate client object data on the server

Netscape the *communicator*
calculate or gyrate
command & view a better *Webscape*
our true virtual escape!

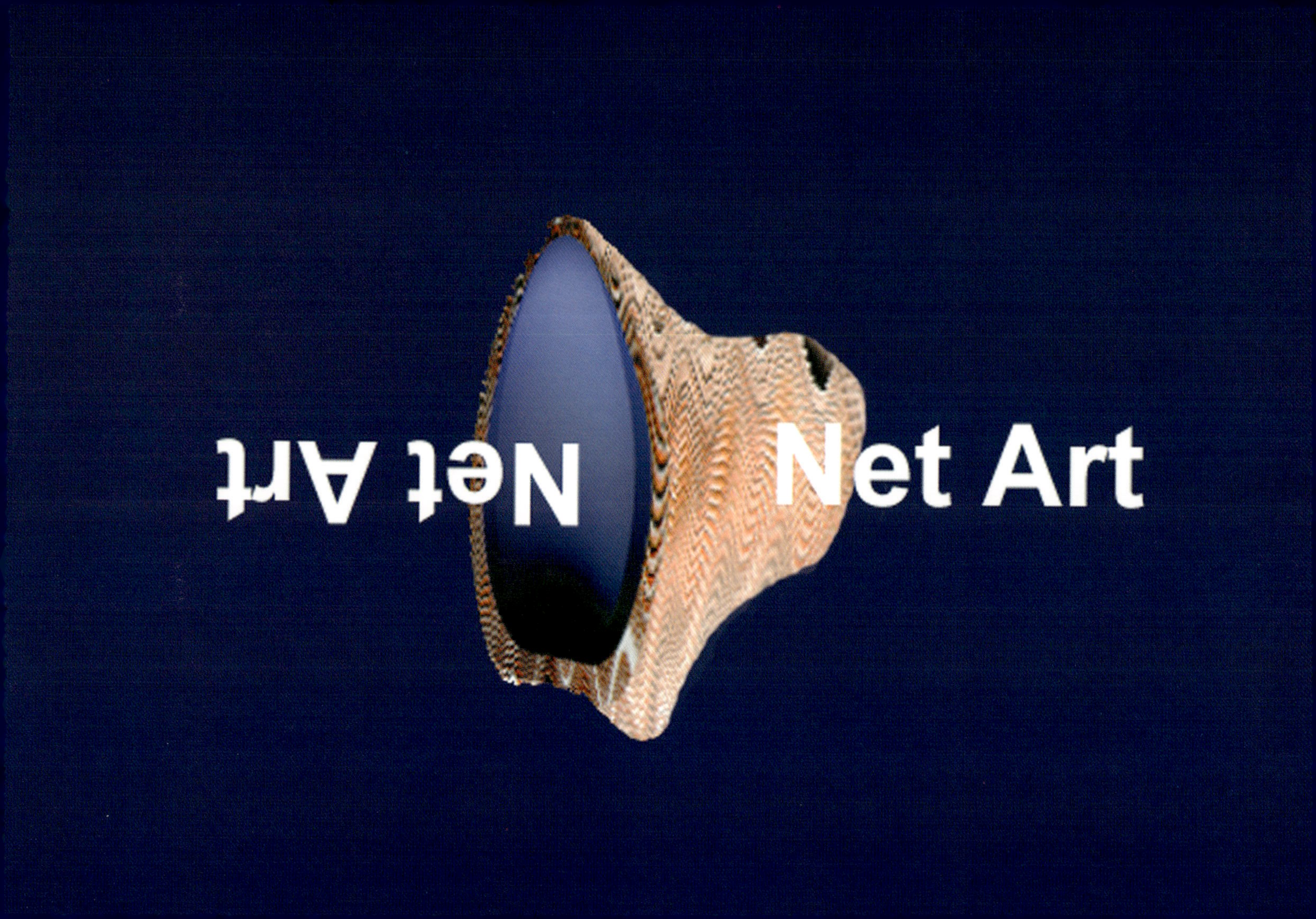

Net Art

chaos in aesthetics
aesthetics in chaos

AESTHETICS ON THE NET

Sep 18, 97

aesthetics on the Net
are both static & images animate

noise & sound
voice & music congregate

as they aggregate
they activate & agitate

they enter & exit
real life emulate
live & then fade

aesthetics on the Net
are both icons & graphics consummate

clicks & links
text & video congregate

as they accumulate
they activate & irritate

net objects
real life eject & reject
live & then decay

aesthetics on the Net
rather, it is:
net on aesthetics?

if there's texture In Real Life
there's texture In Digital Life
the artist texture mapped IRL to IDL
hence giving Digital Palette a new lease of life!

GRAPHICALLY SPEAKING

Sep 14, 95

graphics enhances *Web pages*, hopefully
but we must do it correctly
think of those poor guys far away thoughtfully

resolution, always set to *72 dpi*, painfully
resample it to pixels desirably
covert to 256-color *gifs* or *jpgs* accordingly

when you're greedy
filesize becomes large naturally
aesthetics, your only concern, maybe?
performance will outspeak frequently
though not necessarily
nevertheless, it defeats the whole purpose graphically

for those who can't wait patiently
the *Stop* button is always ready
to kill a transmission instantly

for those who survive *Net glitch* genetically
patience will soon be drained when overstretched periodically
thus, a text version may help wisely?

ANIMATE OR CHECKMATE?

Sep 23, 97

since that day: have pencil will animate
blades or plates,
so long as they're *objects*, they must be animated
otherwise, they're considered as relics
& you'll be eliminated

to animate
begin with a concept, not straight
funds, you persuade
2D or 3D animation software great
fastest hardware must parade, else simply wait!

if you've a concept straight
don't waste time,
don't even try, babe
stay awake, generate good ideas
else wait...

if you've a concept that raids
creates models & textures, real & fake
gears up camera & lighting, don't be afraid
animation paths make & rendering begins to take place
now, try and imagine, you're standing on the lake!

but, if the hardware does not cooperate
& the animators have their periodic breaks
go for a holiday & shake their legs
when you're next awake,
someone else's news checkmates!

art in technology
technology in art

inter@©tive

Sep 23, 97

inter@©tive
what is it?

it's **in**
it's **active**
therefore it's **inter@©tive**

any word

if you can't define
it's fine

otherwise, check the dictionary
it's a better find

if you don't have a dictionary
buy!

otherwise just quit
& say goodbye

seriously, any word

if you can't define
don't pretend, don't redefine
worst still,
influence the passer-by

....

so what's **inter@©tive** @rt?
it's **in**
it's **active**
it's **art**
therefore it's **inter@©tive** @rt

any meaning

if you can get by an ***authorities' eyes***
& if you can convince the guy to buy
you've got a finger in the pie!

otherwise, sit back
don't cry & sigh!

any meaning

if you can't correctly define
don't give up, broadcast & redefine
shake & break the pie!

this is how
even when a fly lands on a *TV screen*
with or without moving
fools those that standby!

TV screen = video art

PIX GENRATES PIX!

Sep 19, 97

if you have a pix
on the Web, you affix
next, organize & get an event date fixed
broadcast the event & upload the pix

worldwide Web artists will get their eyes fixed
retrieve the pix
morph or tween, mutilate the pix
generate & upload the new pix

same or more artists get these newly generated pix
more tricks, more tweaks
& more new pix
again upload this new generation of pix

processes repeat
all you need is a Perl Script
so that the right *child-pix* would be affixed
to the right *mum-|-dad-pix*

more child & cousin pix, more great-grand-child-pix...
no in-breed, but soon you'll see an ocean of pix
a genealogy of pix
indeed, the picks & spikes of the original pix!

pix = noun, both singular & plural, coined by the author

| = or in mathematics

VRML

Apr 3, 96

you can now fly over your neighbors' roof top as a passer-by
or peep into his house on a drop-by
drive out the garage just before you say good-bye
whether you parachute up or drive down

never before, you can change gravity
while you steer & see
if excitement make arenalin greed

you can choose to collide and greet
on this great VRML beat
while you stand firm on your own two feet
all are built on shapes 3-D primitive
based on basic geometry

slice or glue
splice or sweep

its potentialities
extend to researching
medical impossibilities
architectural & engineering modeling revolutionary

virtual reality
when exacting to fulfil non-unque properties
is boring
thus depleting artistic creativity!

VIRTUAL or REALITY?

Sep 19, 97

must virtual reality be made a reality?
beam up to life size to tease & please
to make a show from 0 to 180 degrees
from 180 degrees to 360 degrees

yes, one must forgive
the virtuality of reality
& understand the importance of
the vitality of virtuality

The reality of virtuality
is to dispose virtuality
& live in the realm of reality
& disguise off as true reality

build a shell for virtual reality
plug in the same (best possible) structure in different localities
surface texturing aesthetically
thus nativity with different nationalities

playing life-size puppetry in such *captivity*
is distancing reality
perhaps even gimmicky
thus it is no miracle in near visibility

virtual reality
is a technology brave out of dexterity
with quality by Mark Peces, the visionary

ALL THINGS NICE & PIXEL-DEEP!

Mar 28, 96

if beauty is skin deep
then digital beauty is pixel deep

when ideas bleep
the mouse clicks
16.7 million colors & 24-bit color palette dip
pixels instantly reap

cut & paste
morph & warp
rotate or flip
pix is just within a click

tween or tweak, apply masks or filter tricks
be it image solarized, blurred or crisp
formatted in pct, bmp or tiff
all must be resampled to 256 color & 8-bit
jpeg or gif
to be beefed up on TCP/IP -- the Internet beat

however, any high resolution or subtle graphics
just cannot withstand this scaled-down beat
hence poor quality image creep
bandwidth, the cyber-weep!

WHO IS THE OWNER?

Sep 19, 97

in the real world
when an artwork is scanned
it's transformed into pixels

in a digital world
when an artwork was originally conceived by technology
it's represented by pixels

whether these pixels are being printed,
transmitted or remain status quo
who's the owner?

just like in the real world
steganographicize your images
& time stamp your creation

to identify the owner & the one who contributes
record the image *contents attributes*
discriminate access & use rights to prevent disputes

however, can the metadata withstand
the change of resolution, perspective views or rotation
image resampling, flipping or cropping?

can the medata withstand re-palatettizing
image be blurred or sharpening
trace or embossing?

can the image be cut & paste, morph & tween
print & again transmit or transfer in between?
or even stand subsequent washing machine handling?

this will depend on

the art & science of water-marking
& the awareness of legal whacking
or rather the love of sharing & not sharing?

steganographicize, verb of steganography, coined by the author

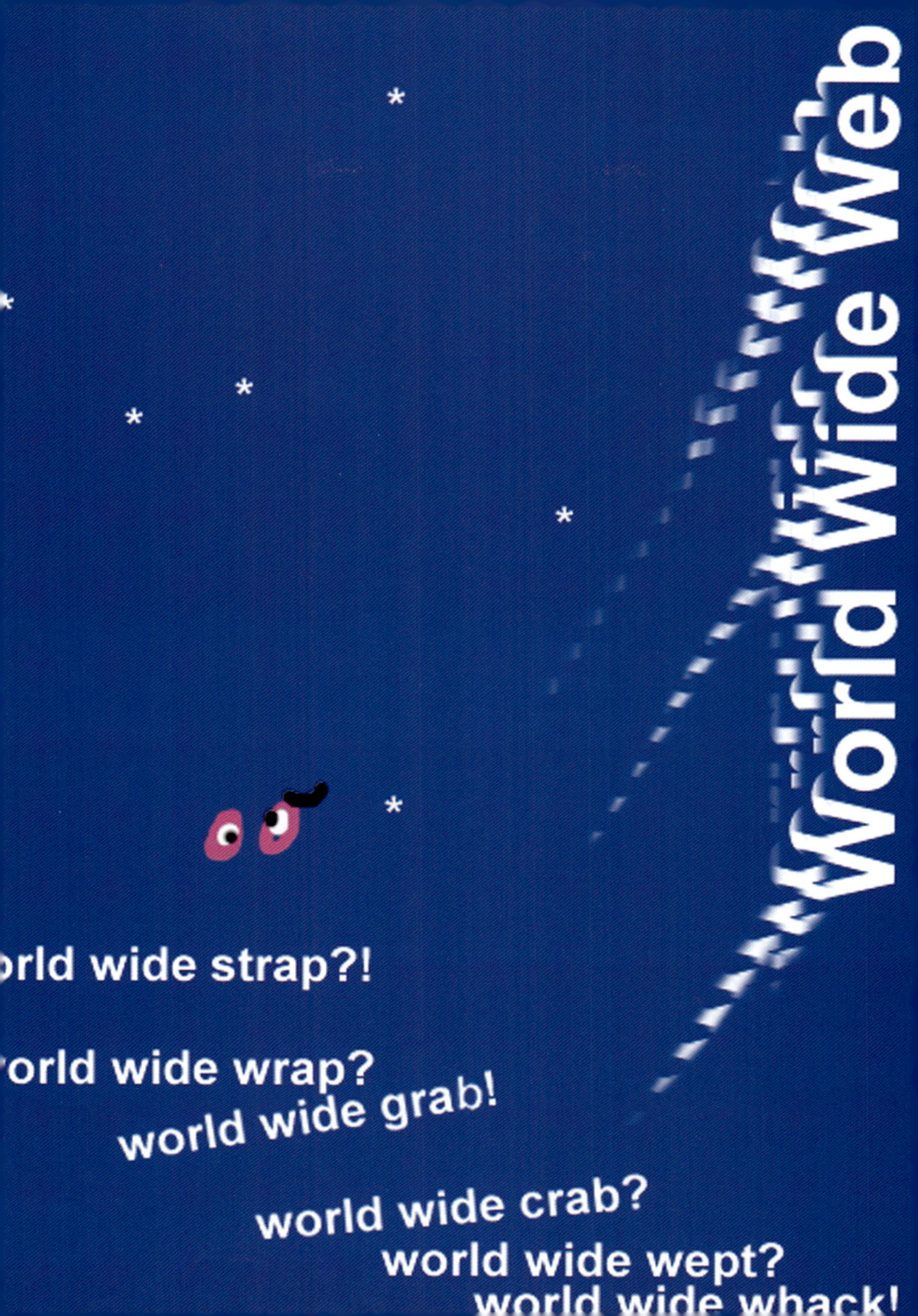
World Wide Web
orld wide strap?!
orld wide wrap?
world wide grab!
world wide crab?
world wide wept?
world wide whack!

the fusion of the Web
is

to diffuse
to infuse
to confuse
to refuse

LINKS

Mar 7, 97

jinx & goodies
all can be reached easily
by clicks on links on very Web page & locality

by choice or curiosity
you click to be linked
& you link to be clicked

click on links of the links
& be linked by subsequent links
hence reaching out to more jinx & goodies

from pole to pole
from links to links
these are links
links by cities & countries
links that unveil events & connectivity
links that unveil availability & potentialities

from pole to pole
from links to links
these are links
links to unveil personalities
from crooks to dignitaries
from mafia to celebrities

from pole to pole
from links to links
these are links
links to condemn & links to redeem
links to understand cause & consequences

from foreground to background
from aboveground to underground
it's links that unveil the truth & ugliness

from hype to supremacy
from surprises to shocking discrepancies
from capabilities to cavities
from intimidation to intimacy
it's links that generate irregularities
it's links that degenerate regularities

is it links that transmit & permeate
diversity in unity
or unity in diversity?

links, assets or liabilities?

LIFE IN THE FAST LANE

Sep 18, 96

browsing began in the University of Illinois camp
later, along came the Netscape Nagivator lane
where Marc Andressen was brought to instant fame
& Bill Gates slept in insomnia pain

while Netscape naviGates
Gates explored the virgin highway plain

he explored & crept, peeped & pricked
sped along, browsed & stripped
hence opening a newer lane, perhaps war?
indeed, the Microsoft Explorer is to be blamed

while Netscape stays in Hyper Text Marc up Lane
& Netshaped into the HTML 3.0 extended terrain
it again naviGates Bill's Winner's brain
highway taxed and Navigator must be tamed

while Java continues to script
Microsoft bills paid & border extended
heads beveled, color shaded, mouse up and picture framed
ActiveX crawled & jumped out of its bibs and cribs

while users could now ride & race side-by-side
on the newer & faster freeway lane
compare and contrast to bet on a winning game
users are the one who clearly gain

from Mosaic to JavaScript
while users eagerly wait for less plug ins & better trips:

he who naviGates must now gauge & re-engage
what to strategize next & how to close the Gates

he who opens the Gates, must now re-think & re-calculate
once security slips through the wide-opened Gates
users will not only be aggravated but close the Gates
steer away, re-meditate & investi-Gates

if you choose to race in these high-tech lanes
life is such that, it does not matter what or who innovates
it is what has been explored, who steers ahead & navigates
while users unceasingly watch, wait & again upgrade!

as obsolescence is constantly on the shrink
if arrogance is the key to "m o u s e   o v e r"
then success will depend on how far you can sustain
& how fast others can make the grade!

HOME AWAY FROM HOME

Sep 7, 95

this is a page called *home*
a home away from home
it's often standalone
& you can call it your own

this is a home
same language
but different personality sewn
& your identity shown

this is a place
people can visit any home
with a standing invitation
& without making their presence known

but their digital fingerprints alone
can make their presence known

LIBRARIES WITHOUT WALL

Sep 15, 95

this is a library
without wall & floor
without shelves on & off the floor

manuals, dictionaries & thesaurus
in all subjects & all languages
are placed all over the ocean floor

no librarians, no indexes
no classification, of course
no catalogs available in this cyber floor

as it grows wildly on the floor
transmission speed enthralls
classification phobia also draws

meanwhile, eavesdropping across the wall
cut & paste on the floor, duplicates pour
but everyone uses it, everyone adores

nobody controls
hence it exhibits no flaws
that's why, the *Web* is free for all!

aNYTH I n G GOES

Sep 13, 95

this is a place
where anything goes
as long as the author achieves his goals

be it terrorist handbooks or Playboy centerfolds
what's unconstitutional & whatever pursuits personal
you can simply throw

whatever you grow
as long as no authorities know
you won't get told

gifs or *jpgs*,
be it obscene or indecent
lewd or lascivious
opinions, timid or bold
gracious or ungracious
entering the *Net*, is gaining a foothold

to the doer
everyone does it
so, why withhold?

to the viewer
everyone reads it
so, download while the gate remain unclosed!

browse?

Oct 21, 97

this is where you begin to browse
as you click
you will either raise your eye-brow or frown
or simply say "*wow*"
or venture into yet another link & again browse

as you continue to click
there will be endless *"wow"s* or frowns
before you frolic into further clicks
your digital clock silently ticks
& every micro second counts

this is when you begin to realize
you're actually deadline bound
in this space you drown
or perhaps
a new paradize found?

World Wide Web u n l i m i t e d

Sep 9, 95

universities and labs
museums, shops and corporate
auto drop on your lap
without having to move a step
or swim a lap

bookmark & classify by subject
by links or by object
hypertext well-sketched on your notepad

you can now study step by step
tour the virtual world
be guided by the weather map
all at a click on the World Wide Web

encyclopedias, thesaurus and proverbs
freeware on the Net
without picking up the tab

office around the world
opportunities abundant

pixel dab
exhibit art on the Web

ribeyes, fillets and mignons
eat & digest on the Web

moral principles
indeed, exhibit the same pattern
in real & cyber world!

WORLD WIPE WEPT

Apr 3, 96

whether you're a start-up in California
or a multi-national corporate in New York quadria

be it technologies or crabs
you make your presence felt on the *World Wide Web*

& hence the beginning of
The Webstock generation euphoria

however, in less than 2 years
just last week, when a 30 million dollar flat loss swept
the Times Board of Director wept
hence calling the shutdown of a vision called Web

as netizens surf across the red sea oceania
they'll begin to see some of these high profile Web sites
will soon be out of sight
& lose their mighty might

as the sclerotic arteries clog the pipes
economics of profit will be the first on sight
whatever redundancy & arrogance aside
venture capitalists must first put the bottom line right

nevertheless, don't get frightened, don't lose sight
netizens world wide wept aside
all must be prepared to fight
re-formulate & get your *Netstrategies* right!

Vitamin A to Z
only if you click
then you'll see

but don't be greedy
not all *sotongs*, prawns & crabs
are cholesterol free and must grab

once appetite whet
experiences, no way you can forget
it's hard to take your eyes off the Web!

sotong = cuttlefish, Malay

WORLD WIDE WHACK

Oct 7, 96

this is a place
where voices meet

they can scream or screech
to exercise the freedom of speech

whatever they believe in, they preach
to as many as they can reach

topics are rich
whatever they don't believe in,
your lips they'll try to seal & stitch
at the same time raising their own pitch

chats are clipped, threaded & sandwiched
all parties can talk concurrently
without a hitch

contents can be switched
anyone can join in at anytime, without a glitch
but anytime, chats can go into a ditch!

while information is exchanged & gossip enriched
threads are threats or treats?!

ODE to TRAVEL

Nov 28, 95

to travel
is to marvel

marvel?

yes, it's remarkable
to be able to travel
thru' time
& thru' space

these days,
not only do we travel
thru' real space
but also, space cyber

via dialups, T1 & T3
ATM & ISDN
& the latest GlobeSpan
@ an amazing 6 Mbit per second

to travel
is to wonder

wonder?

yes, it's tangible
to be able to travel
thru' a real space model
but also, reality virtual

via media digital
in text & images
in voice & motion pictures
in gigs over copper wires

to travel
is to ponder

ponder?

yes, it's applaudable
to be able to savor
the different flavors
of the notion travel

via a stopover
thru' Springtij Festival
in this space cyber
1995, from 6 to 17 of December

Visit the Ode to Travel Web Site @:
http;//www.lhham.com.sg/exhibit2

WELCOME TO LIN HSIN HSIN ART MUSEUM

Apr 19, 95

when you wake up, we take a rest
whether you're from the East or West
we invite you to sign in & be our guest
we're open 7 days a week & 24 hours a day
in this art-friendly cyberspace

we encourage you to mouse & browse
feel free to walk around
we're sure something that you like can be found
if you're our visitor the 20th time round
you'll become our Friends of Museum in & out of town

there is so much to do in this Museum
we invite you to take it all in, absolutely without traffic jams
visit the Oil on Canvas, the Oil on Paper
& the Works on paper galleries
find out more about paper in our workshop

if you miss the sun, visit the Sun in the Studio
learn about the artist's view on man & woman
savor her humor & her inspiration
from art to science, from technology to poetry
she treats them all as equals

if you like to meet our artist + information technologist + poet
you can watch a video in our Theatrette
listen to music when you put on your headset
you can rest your "*feet*", sip coffee & lick ice-cream
in our *suppa dupa* cafe, while looking out at our tropical garden

be sure to visit our toilet, learn about toilet culture
while you perform the calls of nature
even as you rush to flush, you can quietly trust
in paintings you nurture
in poems you mature

before you depart
don't forget to visit our gift shop
sample an anthology or catalog
shop for posters or postcards
or even mail order an original work of art

as there is so much to do in this Museum
visit us as regularly as you wish
drop a line, we would like to hear what you say
look out for coming attractions
on this site -- the truly artistic lane

visit the Lin Hsin Hsin Art Museum on the Web
http;//www.lhham.com.sg

OIL ON CANVAS GALLERY

Apr 19, 95

be it stars or planets
sound of music or echo
where water flows
paints thrown

from time to time
inspired by man & woman
books & nature
images unfold

travels abroad
memoirs & records
merits accord
& canvas grows

'tis a gallery
where art, science & technology
paintings & poetry
live together as a family

http;//www.lhham.com.sg/ooc1.html

OIL ON PAPER GALLERY

Apr 19, 95

pastel or pencil
ink wash or watercolor
all can be painted by using oil the color

visit this gallery
to dispel opacity
to witness translucency

visit this gallery
to discard legacy
to understand oil miraculously

from rigidity to flexibility
this is the gallery
with true originality

http;//www.lhham.com.sg/oop1.html

SUN IN THE STUDIO GALLERY

Apr 19, 95

pushed by the waves
shaped by the tides
as I gaze into the paradise of sun & sky
on paper, with fluorescent paints I speak my mind

in the fullness of time
I make time
& I mark time
between tides

moments I cherish
moments I relish
& moments that vanish
are all reflections of time

rings of time
swatched against the sun
a wink & a whip
is time for a siesta

'tis the time to meditate
time to think
time for reflection
& time for recollection

thru' this hour glass
in my studio
I share with you
the melting sun
perhaps, for you
the midnite sun
as the sun rises
or reposes

http;//www.lhham.com.sg/sun1.html

Concerto for shit

May 15, 97

this is a concerto
composed & played impromptu,
or written for a recital

it's a concerto
to be performed only in solo
alone in a chamber with water

this is a concerto
where the soloist is the composer
the composer is also the soloist

this is a concerto
where the composer
is also the audience & conductor

never before,
a concerto is composed & played simultaneously
while sitting on a "stool" composing "*stool*"

never before,
the *composer-cum-player*
can compose & play while reading a newspaper
talking, watching TV or listening to the radio
while fingers playing on a keyboard digital
indeed, a *one-person-orchestra* making music chamber

this is a concerto
written for trumpet or trombone
French horn, cello or piano

....

it's a concerto
accompanied by *fugato*
or even *oratorio*

this is a concerto
to be played in tempo *largo* or *allegro*
but often, *rubato* in *moderato*

it's a concerto with three movements
first, the *allegro moderato*
followed by *adagio* & end in *presto*

this is a concerto
with resonance in *legato*
or yodel in *staccato*

as the composer twitters & quivers
it performs & morphs itself into a *rondo*
with memories & come back to Sorrento

as the composer fiddles
it drops & plucks in *pizzicato*
it whistles & warbles

when the performance is over
it flushes itself into water
& applause into *concertante water*

when the performance is over
the composer records the manuscript
by brushing it onto tissue paper

this is a concerto
that plays with a tone color
brown or ochre
any variations in color
need to be examined by a doctor
for its color & odour

any detection of *moto perpetuo*
needs to be examined by a doctor
to analyze the composer's diet *capriccio*

this is a concerto
obbligator, composed as a daily *arpeggio*
& performed as a *toccata*

this is a concerto
without manuscript & *intermezzo*
complete with senses overflow!

visit the Concerto for Shit Web Site @:
http;//www.lhham.com.sg/exhibit

it's shit that rules!

May 15, 97

be it chicken or duck
peacock or cuckoo
elephant or *moo-moo*

everyday, everywhere,
every hour, minute & second
birds with feathers
or animals domestic or in the desert
will stand alone or gather
somewhere, somehow their bowels will move

be they geniuses or fools
architects or engineers
artists or administrators

everyday, everywhere,
every hour, minute & second
babies that chuckle or adult marathon runners
those with pretty face or others with tattoos
will sit, squat or stoop
to ensure their bowels move

thus
it's shit on the road or roof
on the rock or in the pool

it's shit in the loo
loos in the plane or train
loos in the offices or museums

everyday, everywhere,
every hour, minute & second
the hypocrites & the squares
those that smile or swear
will shit while they talk
read, write or sock

thus
it's shit in the mind
or queuing in the spool

thus
it's shit by the geeks & nerds
& shit by the unkind

be it shit from the bowel
or shit from the mouth
it's shit that benefits or destroys mankind

whatever its shape & size
the diet that we choose
determines the color of the stool

whatever its shape & size
the lifestyle that we choose
determines the amount of shit we loose

as we need to feel good & cool
it's shit that makes us cool
it's shit that make us loose

we shit when we loose our cool
or we shit to make ourselves like a fool?
it's shit that make us hang loose

with shit we fertilize plants & irrigate
without shit we'll vegetate & fade
rot & *face the date*

on shit we bark & meow
with shit we paint & draw
mold & sculpt, huts with walls & roofs

on shit the germs archived & grew
germinate, flee & fly
by the millions & permeates disease like flu

it's shit that does us well
our health & wealth
on shit we live & dwell

it's shit that proves & tells the truth
it's shit that approves or disapproves
ultimately, it's shit that rules!

visit the It's Shit that Rules! Shit Web Site @:
http;//www.lhham.com.sg/exhibit/shitr

SHIT PLEDGE

May 15, 97

We are the shit of the world.
Regardless of your race or religion,
cultural heritage or financial mortgage,
male or female, marital status,
age or vintage,
we look n feel the same!
Same color, same hues n shades,
same smell and same shape.

Whether you are a six footer or underweight,
pretty or configuration without shape,
we are the same,
yesterday, today and forever,
same facets and same packaging.

Whether you use technology on the bleeding edge
or you still use your fingers to count a page,
we are generated by the same input device,
processed by the same generator,
& printed by the same output device.

Whether you are located
on the pole or the equator,
we say hello to you day by day,
in health or in weakness,
till death do us part!

STATE OF THE NET

May 4, 97

from love to chats
from software to pets
all can be heard & seen on the Net

once your appetite has been whet
you'll be looking for the spread
& be stuck with clicks on the mouse pad

be it facts or crab
whatever you wish to see & get
it's quite likely on the Net

be it fiction or a sketch
whatever you dare to bet
clasp & strap yourself on the Net

whatever the state of the Net
you must cast & broadcast on the Net
so that you can sow & reap on & off the Net

whatever the state of the Net
beware of the loopholes & catch
so that you will not get reaped off on the Net

whatever the state of the Net
take care of your properties on the Net
so that others can't claim what they've tapped

chaos or order,
this is the state of the Net
the scene that we've set
on which, the future we've bet!

visit the State of the Net Web Site @:
http;//www.lhham.com.sg/exhibit3

BACK TO FUTURE

Feb 14, 97

from hieroglyphics to graphics
it's back to basics
it's pixels that speak

art as exhibits
has transformed

from real to unreal
from unreal to surreal
from surreal to superreal

art as exhibits
has exit

from its physical space
into a virtual space
at a sonic pace

from oils to acrylics
from sculptures to ceramics
from sketches to images sleek

art as exhibits
all are reduced to pixel deep

from cartoon to lyrics
from kitsch to satiric
all are recycled bits

looking at art as exhibits
are pixels still as exotic?

from artists unknown to famous picks
from occasional to serious geeks
in virtual space they cohabit

as pixels squeak
is experiencing art still as esoteric?

as the meaning of art we continuously seek
we ask, what then is aesthetics?

*visit the **Back to Future** Web Site @:*
http;//www.lhham.com.sg/exhibit4

NetFUTURE

WHAT'S THE DIFFERENCE?

Sep 18, 97

In the cyber world

swear by instant noodles
& caramel water

whispers & murmurs
giggles & jingles
love & murders
co-exist & co-mingle
pixels & pixels
facts & fables
mimic & mirror
clicks & links without intervals

life in the cyber jungle
no commanders
all fighters
bits recycle
ends get further

no wrinkles!

In the real world

swear by food that puzzles
business & swindlers

whispers & murmurs
giggles & jingles
love & murders
co-exist & co-mingle
people & non-people
facts & fables
mimic & mirror
drink, eat & shit with intervals

life in the real jungle
commanders, commandos & fighters
sleep, sit, walk, drive & bicycle
life physical, worries metal
ends get nearer

generate wrinkles!

ARE WE HUMAN?

Sep 19, 97

it was first the video conferencing
then *See You See Me*
& Netcasting

with all the cyber protocols
cybernoids should be exercising
click to meeting & meet by clicking

but, we are still attending real world conferencing
experiencing flights & hotels overbooking
restaurants, under catering

we are still constructing real world buildings
museums & offices skylining
transporters in the sky or on the highway queuing

indeed, the real world is insufficient for our living
cyber world is insufficient for our feelings
combining them is certainly better for communicating

indeed, we desire to advance real world engineering
to enhance our cyber world living
so that we can escape into the cyber world
to make a living
to readily express our feelings
& to e-commercing!

is this forgiving or contradicting?

ALL THINGS BIOLOGICAL

Oct 3, 96

in this space cyber
though you can do wonders
you can't do things basic & normal
you can't manage all things biological

for example
you can't smell, let alone drink & eat
hence you can't *pee* & *poo*
shit & sleep

though you can still sneeze without a hiss
when you need to hug or seal a kiss
you'll dearly miss a kiss
& be bugged by a hug

nevertheless,
you can grow a plant tomato
& bear the fruit as cucumber
for that matter, stimulate the fig to flower

you can rear a cat
& grow it as a tiger
you can conceive a baby boy embryo
& change the gender

though you can reorder your memory
rectify your stomach disorder
you can't grow thinner or fatter
you can never fulfill your functions biological

....

in this space cyber
it changes your body temperature
alters your blood pressure
pleasure or pressure, it makes me wonder!

CYBERMEDICS

Sep 19, 97

if a cyberbody becomes ill
what can we prescribe?

first, we must describe
what's a cyberbody

is it logically physical
or physically logical?

then we must transcribe
what's cyber sickness

is it sickness in cyberspace
or is cyberspace sick?

afterwhich, we can then prescribe
cyber medication

for the real body in the cyber world
or cyberbody in the cyber world

or real body in the real world
or cyberbody in the real world?

this will depend on
whether the real body is controlling the cyber world
or the cyber world is controlling the real body!

EXAMINE FROM AFAR

Sep 19, 97

video in the internals of a real body
have the real body's medical history ready

distant surgeons can co-examine the real body from afar
heart or intestines, pulse rates or blood pressure data
all can be read from the toolbar
while sitting on an *office-sofa*

medical advice and operating commands from afar
can thus be communicated instantly
to say the least, on par with that performed locally
or even better, with special remote plugins

video out to display the results proudly
have the real body as a testimony

this will be a situation *win-win*
so long as there are no transmission hang-ups & side gagas
otherwise, the real body may soon see stars
& medical experts will soon disappear on the screen from afar
thus medical advisory becomes adversity
& technology gets bad publicity!

CYBER OLYMPICS?

Sep 19, 97

be it swimming or cycling
marathon or decathlon
we can emulate & compete on the Net

be it netball or basketball
softball or baseball
we can play them all

the day will come
when we can scan in our *cyberbody*
& connect *every(cyber)body*

we'll compete with our *cyberbodies*
with our technological capabilities
to see who has a stronger *cyberbody*

we will then conclude, at least in theory
not who has stronger physical muscles
but who has better *cybermuscles*

as we begin on the mark
we are ready to go
ping & get *ping*

as the *cybereferee* signals
bits gather for rehearsals
bandwidth, the conductor

....

transmittal or fatal
winner or loser?
no *cyberbod*y answers!

CYBER-LY PUNCTURED

Sep 19, 97

if the real body is controlled by
neurons, dandrites & nerves
then map the x,y,z co-ordinates oeuvres
to a server

if the real body nerves are governing the aches & pains
then correcting the mirco positioning of these nerves
will converse with these aches & pains
& supress or reverse these aches & pains

if we can acupuncture our cyber body
then *cycupuncture* our real body
readily & logically
would be a possibility or an impossibility?

if we can scan in the x-y-z nerve oeuvre procedurally
then assign an IP address to every acupuncture point accordingly
connect electrodes to the our real body
all electrodes are conected to server(s) accordingly

ping the desired IP address
our real body nerve would be pulled accordingly
but adjusting it correctly or incorrectly
will depend on the accuracy & sensitivity of our cyber body

if we can *cyupuncture* our real body accurately
that will be yet another way of using technology
to medicate or meditate momentarily
such is a real world fantasy or cyber world fallacy?

E-MAIL ONLY!

Sep 19, 97

imagine,
this is a day absolutely without traffic
no real world (*wide*) traffic
no cyber world (*wide*) traffic

wouldn't this be terrific?

imagine,
this is a day where everybody remains stationary
at his/her x,y,z coordinates respectively
though there's electricity
let there be no *electrovity*

would the Web be weeping?

imagine,
sometime somewhere & somebody is brave
to be sent to a cave
& must live there alone for 365 days
with daily supply of electricity, food & drinks in the cave

however, the condition is:
s/he can sleep, eat, shit, sit, stand & walk
but not be allowed to talk
s/he can be connected to the Net, gawk & hoard
play, listen (*only*) to music & e-mail to anyone *on-board*
there will be response for every e-mail

can somebody continuously *not talk* for 365 days?

e-mail: poet@lhham.com.sg

electrovity = any activity powered by electricity directly or indirectly

don't talk = absolute silence in any audible format

THE DAY WILL COME...

Sep 9, 95

we now *talk*, *read* & *write*
in a space that has taken over our human rights

we do face difficulties
in using our human faculties in their original light

our lips are tight
we "*talk*" as in *write*

*read*ing off the screen is what delights
we exercise all ten fingers to *write*

the day will come
we bite to execise our jaws
we hold chopsticks to test our ability to draw
we point at tree leaves to count to four...

as we continue to "*type to jog*"
we again see cyber traffic lights
jam across the conduits & pipes
perhaps, it's time to fly a kite?

EPILOG

this may be your first time, techies
you hear my voice on hardcopy
or on cyber territories
enjoy *new-p-bies*!

this is my 5th volume, incidentally
with the exception of the 3rd on information technology
the 1st, 2nd & 4th being poetry
on lifestyle & issues expressed *philosophically*

thanks to those who inspire me daily
for they contribute to my perspiration continuously
so I can then talk relentlessly
till then, you'll hear from me again, definitely

What reviewers say...

*On "**In Bytes we Travel**", 1997*

"Hsin Hsin scores again with her collection of techno poems. Her intimate knowledge of the computer industry has enabled her to give a humourous and jovial twist to issues surrounding the IT world. A must-read for techno freaks."

Grace Chng, Editor, Computer Times, Singapore

"This piece of literature shows a keen understanding of both technology and art. It is a good marriage between the logical world of technology and poety of the art form. I strongly recommend everyone to read this."

Lim Chin Hu, Country Director, Sun Microsystems Pte Ltd, Singapore

"Refreshing and very interesting that covers the lighter side of Internet."

Ong Teng Beng, Country Manager, Malaysia

"She interlaces words into funny, witty and thoughtful poems. They tinkle in the reader's mind like a summer breeze gently shaking wind-bells."

Lin Hsin Hsin is also romantizing the CPU. She is so fond of computers that she delivered a full book of verses chanting - and duly teasing - the machine of her daily profession."

Lionel Dorset, Journalist, Multimedia Japon, Tokyo, Japan

*On "**Sunny Side Up**", 1994*

"In this, her fourth book of poetry, Lin Hsin Hsin ranges across a spectrum of human experience: from cosmetic surgery to charity; art to information technology; shopping to male-female relationships. her poems are accessible to all, have a lot to say about the contemporary society, and comprise a variety of charming juxtapositions of words and sounds that surprise, startle, and delight"

John Knagg, Director, The Bristish Council, Singapore

*On "**Love @ 1st Byte**", 1992*

"Some of Hsin Hsin's poems are educational, presenting better-than textbook definitions and explanations for the difficult computer jargon and acronyms. In fact, I found the entire book was extremely entertaining, full of WYSIWYG humor & wit. The cartoons, the paintings and the melodic verses make it truly a Multimedia Poetry Book."

Dr Ifay Chang, Manager, Applications Institute, IBM Yorktown Heights, USA

*On "**from time to time**", 1991*

"Lin Hsin Hsin's new collection is a kaleidscope of entertaining poetry. She draws her themes from both the serious and the trivial; you will find powerful evocations of the horrors of the Gulf War side by side with whimisical commentories on lucky draws and closing-down sales. Yet whatever the subject you are drawn in by the deceptive simplicity of the lanuage and the form of the poems. Once inside, you find yourself confronted by a new and unusual viewpoint, often expressed with a gentle humour, which obliges you to re-examine your own assumptions on matters great and small."

Graham Millington, Director, The Bristish Council, Singapore

*On "**take a word for a walk**", 1989*

"In these apparently effortless poems there is therefore a range of subtleties streaks of brightness and shadow and great variation of colours which move the reader no less than those who look properly at the apparently light abstractions of her paintings."

David Howell, Director, The Bristish Council, Singapore

What readers say...

"Poems such as "*In the name of Research*", " *To Scan*" and "*Life in the fast Lane*" reflect "*Life in the reality*" one that is still full of human weaknesses. Though so close to the reality, Lin Hsin Hsin has the intelligence and courage to describe such weakness wittily.

Dr Mouw-Ching Tjiok, IT Consultant, Böblingen, Germany

"Hsin Hsin is a quintessentially poststructuralist artist plying multiple media and the voices of each medium. By using the Internet to talk about new technologies, she fuses medium and message in a playful combinatoric of commercial and personal. She is hyperactively prolific, mirroring the phenomenal growth of information technology and the effects these technolgies have on the fundamental structures of our humanity."

Miko Matsumura, Java Evangelist, Mountain View, USA

"Hsin Hsin's ***'In Bytes we Travel'*** is like a good serving of antipasti - with much variety to tease the senses. It's quirky and whimsical at times, while offering much food for thought."

Lau Joon Ni, Journalist, Singapore

"Simple to read at first, Hsin2's wry poems contain a kernel of truth that is sure to strike a chord with technophiles"

Bina Maniar-Goh, Writer, Singapore

"Contemporary in approach, easy to digest. A must for poetry lover."

Sim Chwee Hong, Librarian, Singapore

"LHH's poetry on technology brings a refreshing & humorous insight to the world of bits & bytes."

Ng Yu Gin, Technical Architect, Singapore

"Hsin Hsin never stops amazing and amusing me. There is always a n+1 ways of looking at things, which most of us would have missed. I love the *Sun In the Studio Gallery.*"

Dilip Soman, International Product Manager, Boston, USA

"Poetry has always been thought of as being the more lyrical of all written texts but Hsin Hsin has left these traditional boundaries far behind her in her reinvented mode of writing that emerges through the vehicle of her electronic stage.

From word to image from image to word -- Hsin Hsin takes us through digital journeys of Java, html and applets of surprises.

As an electronic artist, I am particularly struck by her dedication tooriginality. As she reveals in her poem on scanning - none of her imagery is plundered from existing sources"

Susan Hazan, Electronic Artist, Jerusalem, Israel

"From Protocol to Nature's Call, '***in bytes we travel***'' makes one wonder, whether life is technology in poetic motion or technology is a collection of life's experiences. Yet whatever it is, this book makes us laugh at ourselves on technology and shortcomings in life."

Kpp, IT Consultant, Singapore

"She only reveals the truth, nothing but the truth, the only difference here is, she expresses them in a humorous way."

Peggy Tan, Systems Analyst, Singapore

"If, as Dryden points out, the sole purpose of poetry is to delight, then Hsin Hsin's new collection is an admirable example of how poetry should be written -- relevant and engaging. She feeds us new perspectives and insights into the confusing world of IT, in the midst of her writings, making us feel satisfied that there is some fun after all in this nerve-wrecking age of computers."

Lau Peet Meng, Graduate Student, Cambridge University, England

"Witty poems tickled my grey cells and livened the day."

Li-Chuan Chong, University Student, London, England

"Hi, I'm a student just surfing the Net in the U.S looking for poems. I felt praised having such a wonderful page. By the way I'm S'porean too!!! Its strange I have to come here to discover home and your talent. Keep up this Auesome work!"

Sunand Razu, University Student, Illinois, USA

"I've put you on my hotlist, LOVED the poem with the links in it. Great idea."

Sheila Barrera, Artist/Writer, USA

"Your poems are humourously "cheem". So profound. ...."

David Wee, Management Consultant, Singapore

"Hsin Hsin writes wonderful poems to capture the essence of everyday chores."

Alvin Shwen, Polytechnic Student, Singapore

"Enjoyed hearing you read your poems at The Forum Readings."

Ai Boon, Teacher, Singapore

"I have 2 of your poetry books... Thank you for those "*words*", as they have helped me express what I was feeling then."

Noraini Nain Sardi, University Student Counsellor, Singapore

"Lin Hsin Hsin's Gedichte sind nicht nur für Computerfreaks und Insider ein Leckerbissen, sondern geben auch dem Anfänger auf recht humorvolle Weise Einblick in die Welt der Informationtechnologie. Trotzdem läßt sich der warnende Zeigefinger nicht übersehen."

Martin W Schmid, Chemical Enginer, Gaimersheim, Germany

The Lin Hsin Hsin Art Museum

The Lin Hsin Hsin Art Museum

The award-winning Lin Hsin Hsin Art Museum was inaugurated in the cybersapce in April 1995. It was first web hosted by the National Computer Board, Singapore, and has been hosted by Sembawang Media since April 1996. Since its inception, the Museum has been visited over 380,000 visitors from 90 countries. The Museum has been awarded by the "*Top 5% of All Web Sites*" by Point Communication, U.S.A. in September 1995 and "*Virtuocity Award*" by VXR Corporation, U.S.A. in July 1996.

The 90 countries are:

Andora, Argentina, Australia, Austria, Bahamas, Bahrain, Belarus, Belgium, Brazil, Brunei Darussalam, Canada, Chile, China, Colombia, Costa Rica, Croatia, Cyprus, Czech Republic, Demark, Dominican Repulic, Ecuador, Egypt, Estonia, Fiji, Finland, France, French Guiana, Germany, Greece, Guatemala, Hong Kong, Hungary, Iceland, India, Indonesia, Ireland, Israel, Italy, Jamaica, Japan, Jordan, Kazakhstan, Kuwait, Latvia, Lebanon, Liechtenstein, Lithuania, Luxembourg, Macau, Malaysia, Malta, Mauritius, Mexico, Nicaragua, Nepal, The Netherlands, New Zealand, Norway, Paraguay, Peru, Philippines, Poland, Portugal, Romania, Russian Federation, Saint Lucia, San Marino, Slovak Republic, Slovenia, South Africa, South Korea, Spain, Sri Lanka, Sweden, Switzerland, Taiwan, Thailand, Trinada & Tobago, Turkey, Ukraine, United Arab Emirates, United Kingdom, United States, USSR, Uruguay, Uzbekistan, Venezuela, Yugoslavia & Singapore

What kind of people write poems?

"Are they highly sensitive beings who with tender sighs and "guilded words" waft away from reality?"

Günter Herburger

"Are they people crazy about words to a high degree, who systematically transform states of linguistic achievement into "allegories of form"?"

Paul Valéry

From the poet's keyboard:

"Me? Neither! ......Just can't stop words from flowing......"

"Just like writing equations: writing poems is an act of purifying the dross of profane."

"Just like composing music: writing poems is enriching the virtuosity of alphabets."

"Just like painting: writing poems is brushing over inter-textured words."

Lin Hsin Hsin

About the Poet

The author **Lin Hsin Hsin,** one of the 200 cyber personalities in "*24 hours in cyberspace*", is an *information technologist, artist* and *poet* based in Singapore. She grew up with numbers, bits and bytes with brushes and paints side-by-side. Besides living in the cyberspace, Hsin Hsin has penned more than a hundred articles published in computer newspapers, international proceedings and journals. Her third book "*Love @ 1st Byte*" is the world first collection of poems on computers, can be found in *amazon.com*. ***"In Bytes we tr@vel"***, is her **fifth** anthology. In Real Life, this award-winning artist and poet has had 15 solos and more than 180 group exhibitions in 49 cities and 20 countries across Asia, Europe, North and South America. You can find out more about her in the award-winning Lin Hsin Hsin Art Museum: *http://www.lhham.com.sg* You can e-mail her in any of the 12 programming languages that she speaks or preferably in English, Chinese, French and Japanese.

Index